MW00527515

From Darwin to Derrida

From Darwin to Derrida

Selfish Genes, Social Selves, and the Meanings of Life

David Haig
foreword by Daniel C. Dennett

The MIT Press
Cambridge, Massachusetts
London, England

This book was set in ITC Stone Serif Std and ITC Stone Sans Std by Toppan Best-set Premedia Limited. Printed and bound in the United States of America.

Library of Congress Cataloging-in-Publication Data

Names: Haig, David, 1958– author.

Title: From Darwin to Derrida : selfish genes, social selves, and the
    meanings of life / David Haig ; foreword by Daniel C. Dennett.
Identifiers: LCCN 2019028387 | ISBN 9780262043786 (hardcover) |
    ISBN 9780262358026 (ebook)
Subjects: LCSH: Natural selection—History.
Classification: LCC QH375 .H35 2020 | DDC 576.8/2—dc23
LC record available at https://lccn.loc.gov/2019028387

10 9 8 7 6 5 4 3 2 1

For my mother, who gave me her seriousness and who, by her choice of my father, gave me his enthusiasm.

Would it not be a far more elegant way of interpreting the two-faced image, to say that Janus and Terminus are the same, and that the one face has reference to beginnings, the other to ends? For one who works ought to have respect to both. For he who in every forthputting of activity does not look back on the beginning, does not look forward to the end. Wherefore it is necessary that prospective intention be connected with retrospective memory. For how shall one find how to finish anything, if he has forgotten what it was which he had begun?

—Augustine of Hippo

Contents

Foreword

Daniel C. Dennett

This joyful book tells the story of how meaning came into existence, and how we ourselves came to be able to make sense of our world. It blithely ignores hostile boundaries and unites philosophy and science, poetry and biochemistry, Shannon's mathematical theory of information and good old-fashioned literary scholarship. What could you possibly learn from Aristotle or Francis Bacon about the dynamics of gene regulation, and what could you possibly learn about literary interpretation from the role of retroviruses in rewiring placental regulatory networks? Do scientists have anything to gain by learning the history of their own fields, and can scholars in the humanities and social scientists strengthen their grip on *their* own fields by learning some of the fascinating details of microbiology? Having aroused the passions of partisans on both side of this great divide, I want now to invite them on a roller-coaster ride of ideas, dexterously drawn by David Haig from an astonishing range of sources, from Laurence Sterne's curious novel *Tristram Shandy* (1759) and Immanuel Kant's *Critique of the Power of Judgment* (1790–1793) through the work of Charles Sanders Peirce and Jacques Derrida to the latest peer-reviewed articles on the intricate machinery of gene transmission, duplication, and expression. The ride is all the more exciting because it is in the form of elegant arguments conducted with calm confidence, unfolding a novel vision of life without fanfare or special pleading, letting the implications speak for themselves.

Philosophers, according to a threadbare stereotype, concern themselves with the old question:

What is the meaning of life?

And "hard" scientists, according to another threadbare stereotype, set that question aside, postponing it indefinitely, while they work on a more respectable set of questions, ultimately *physical* or *mechanical* questions about *how* things work, not *why*:

What is matter?

What is time?

How do molecules do what they do?

How did life arise and how do living things stay alive?

These stereotypes have been baked into powerful presumptions, separating the arts and humanities, with their goals and methods, from the sciences, with their differing (*and opposing*) goals and methods. *Geisteswissenschaften* (literally, sciences of the mind) versus *Naturwissenschaften* (literally, sciences of nature—in the broadest sense, which includes moons and mountains, oxygen and oceans). Lawless narratives to the left, laws of nature to the right.

These traditional dividing lines have been taught for centuries, and are still taught today, but they *have* to be taught, since we all know better! We all know that there are *reasons why* the intricate organs of living things are the way they are—if only we can discern them—and we all know that the meanings we find in our arts and humanities are real phenomena, certainly as real as the meanings scientists find and transmit in their equations. There *must be* a way of putting matter and meaning, mechanism and purpose, causation and information, into a unified perspective.

And for over a century and a half we have understood, at least dimly, that the key to this unification lies in Darwin's dangerous idea of evolution by natural selection. In very different ways this has become common ground for many deep thinkers from Darwin's own day to a growing assembly of biologists, physicists, psychologists, and philosophers in the twenty-first century. Aristotle's Final Cause—the *telos* or *what for?* question—has been absorbed, somehow, by Darwin's proposal that life is designed by a blind, *purposeless* process of differential replication, from which emerge products—living things and their artifacts—which themselves have purposes. But how is this possible? There is still an unresolved tension, epitomized by the oft-quoted exultation by Karl Marx when he discovered Darwin's *Origin of Species*:

Not only is a death blow dealt here for the first time to "teleology" in the natural sciences but their rational meaning is empirically explained. (Marx, letter to LaSalle, 1861)

Does Darwin's idea explain teleology, or explain it away? It does both; it shows how real purposes, real functions can arise out of phenomena that themselves have no purpose, no function. But in the eyes of many, this is a dubious assertion, wishful thinking at best, or self-deception. Aristotle's *telos* is a seductive idea but it smells . . . *divine* and hence *anathema*. Should we puritanically abstain from teleology when doing science (but not when doing literature, history, philosophy, psychology), or has Darwin shown us how we can tame Aristotle's shrew, and make an honest woman of her? (This obstreperous or maybe even offensive way of putting the issue is itself an issue to be considered; the submerged distastes and unarticulated biases of thinkers across the spectrum is exposed in the bright beam of Haig's widely informed attention.)

The details matter, more than one might think—more than I had thought until I got swept up in Haig's consummate explanation of how the pieces all fit together. Many of us think we understand evolutionary theory, if not in every minute particular at least in handy if sometimes sketchy outline, and we have tended to think that this is enough understanding to secure a firm foundation for our extensions of evolutionary thinking beyond gene pools and organisms to minds, cultures, and societies. Haig shows that there is much more of value to be gleaned from the intricacies, if we adopt his careful *adaptationist* perspective, treating all the phenomena of life as susceptible to reverse engineering, seeking the underlying reasons for all the patterns we find in nature. Stephen Jay Gould and Richard Lewontin (1979) famously warned us about the polluting influence of this "Panglossian Paradigm" (where we supposedly live in the best of all possible worlds) but we must cast aside our residual worries about the risks of this "Darwinian paranoia" as some have called it, and let Haig lead us into the strange and enchanting (but also disconcerting) world of strategic genes, mindless competitions between renegades and impostors, team players and sentries, robots made of robots made of robots managing to pilot their enormous vehicles into the future without ever having to recognize or appreciate the reasons why they act as they do. This is the world of selfish genes, brilliantly introduced by Richard Dawkins (1976), explored in even more detail by Haig.

Two informal rules can help us place Haig's project on a useful, if provisional, foundation: Braitenberg's Law and the Warden's Rule. Valentino Braitenberg, a Swiss neuroscientist, published an elegant little book, *Vehicles* (1984), in which he described an increasingly complex collection of imaginary organisms, simpler than bacteria (which are far from simple!) and described how they would behave, singly and in groups. The justification for this methodology was Braitenberg's *Law of Uphill Analysis and Downhill Synthesis*. It is easier, he noted, to predict the behavior of a complex entity you synthesize out of simpler entities—downhill synthesis, a constructivist approach—than to analyze the inner workings of a complex entity whose behavior you observe—uphill analysis. He was right, in the opinion of many, and his little book set in motion a considerable research tradition in robotics and other computer-aided inquiries. But something like Braitenberg's Law is also at work in Haig's analysis of the many ways that information gets captured and moved around in the process of natural selection: start small, and build up.

The *Warden's Rule*, if it *can* happen, it *will* happen (discussed by me in *Freedom Evolves*, 2003, 160–161), is something I picked up from somewhere—I'm sorry to say I have source amnesia for this. It is an improvement on "Murphy's Law," that whatever can go wrong will go wrong. It is an improvement because there are real circumstances where a version of it is true, not merely an amusing expression of pessimism. It is supposedly the maxim of every prison warden. Its rationale is that prisons are full of prisoners with time on their hands and a measure of patience and an inquisitive and competitive nature. They will exhaustively search every aspect of the system in which they are incarcerated looking for a way out, or a way of improving their condition, however arduous the path. How clever are the prisoners? Some are brilliant, others just dogged; if they share their discoveries it doesn't much matter who gets credit for the innovations. Instead of testing them for intelligence, test their environments

for *opportunities*. (In *Elbow Room* [1984], I described the existence
of a bag of jewels in a trashcan a few feet away as a *bare opportunity*—if only I had known about it, I could have been rich,
but I had no information at all that would lead me to check
that trashcan for treasure.) According to the Warden's Rule, even
bare opportunities must be counted, since prisoners have nothing better to do than hunt for them. No trashcans go untested
in a prison.

Bare opportunities abound in the world of evolution, and it
doesn't take intelligence to stumble across them; it just takes lots
of time for variations to arise and get tried. If there is *information*
in an evolutionary environment, a difference that *could* make a
difference to something that could detect or just respond to that
difference, it *will* make a difference (generally—this is a rule of
thumb), a difference that in turn can be amplified by reproduction, making a bigger difference and, in a recursive process, making the detection of that difference more likely in the future, and
so forth. Yes, this is just another way of explaining how natural
selection works, but it works this way *all the way down*, so that
even simple molecular structures can be *usefully seen* to be like
those persistent prisoners, ever delving, ever seeking an advantage that they can exploit to further their own "interests." The
prisoners have a single overarching goal: *Escape!* These agents
have a different goal: *Reproduce!*

When you put these two perspectives together, you get the
methodology that bears such delicious fruit in the work of David
Haig. Braitenberg's Law invites us to work bottom up, starting
with very simple agents, material genes, or even their elements,
and treating them as dead stupid *homunculi*—as stupid as we
can muster—with just one goal: reproduce! As François Jacob

famously put it, "The dream of every cell is to become two cells," but we can apply his vision one level (or more) lower than he did. When we do this, we recognize that individual tokens of a gene type are rather like a gang of brothers in prison. They are kin-altruistic *or* we can consider the siblings as a unit, a *selfish* unit. For many Warden-like purposes, it is easier to consider the opportunities available to a selfish gang. Their competitive nature is not so much an urge driving them from the inside (they are just parts of macromolecules, after all) as a systematic talent for taking advantage of opportunities if and when they arise. The reason we have to look at the *strategic* gene is that these gangs do have to work together. The copies in the germ line—the eggs and sperm—get to escape (and multiply, if they are lucky) thanks to the "efforts" of their brothers in the soma or body.

So these families of little robots are like the simplest Braitenberg vehicles, and thanks to the law of downhill synthesis we can plot the opportunities that might arise, and the effect of a bunch of such vehicles "discovering" them. Where Haig speaks about what a genetic element can "expect" he is talking about the epistemic predicament facing these robots: is there *any* information in their immediate vicinity that they could ("in principle") exploit to their advantage? Are there bare opportunities that they could discover by chance and then act on? Could they "recognize" an opportunity? They could in the only way such a robot can recognize anything, by trying it as a random stab and being "rewarded" by a benefit. If these opportunities are systematically available, then they are part of what the gang can expect.

But even if we grant, for the sake of argument, or as a crutch for our imaginations, that we can interpret such mindless things as agents for some purposes, where does *real* agency kick in? How do we get from molecules to masterminds, from selfish genes to selfish (and altruistic) people? Haig tells us that "conscious intentions are special cases of a pervasive intentionality of living things," and he has a lot to say about how these cases are special and how they might arise. Finding illumination in Adam Smith's often-ignored masterpiece, *The Theory of Moral Sentiments* (1759), Haig constructs a version of morality that is distinctly Darwinian but a far cry from some of the simplistic versions of evolution-based ethical thinking currently in vogue. (In other words, any who have been unimpressed by applications of Darwinian thinking to ethics and political theory in relatively crude recent examples should give Haig a chance to salvage the perspective.) "Do I *feel* you feel I am somebody who can be trusted? Do I *feel* you feel I am somebody who can be exploited?" Haig builds a subtly interactive confection of genes, culture, and reason to provide an explanation of how "Integrity is born out of prudence." Not by a miracle, but step by step, in a gradual series of moves to greater and greater complexity, more degrees of freedom to be controlled.

Evolutionary biology has been blessed with many fine expositors, starting of course with Darwin himself. My admiration and gratitude to those fine authors has been often expressed and will not be repeated here, to save the reader from an honor roll that would either run on for pages or neglect somebody whose work has been particularly illuminating to me. David Haig stands out, even in this illustrious company, as a bountiful source of novel insights, a clarifier—and often, in my opinion, resolver—of controversies, a scholar both scrupulous and playful, who reminds me, as well as anybody alive can remind me, how glorious it is to *think* and *understand*.

References

Marx, Karl. 1861/1942. Letter to Lasalle, London, January 16, 1861. In *Gesamtausgabe*. New York: International Publishers.

Braitenberg, Valentino. 1984. *Vehicles: Experiments in Synthetic Psychology*. Cambridge, MA: MIT Press.

Dawkins, R. 1976. *The Selfish Gene*. Oxford: Oxford University Press.

Dennett, D. C. 1984. *Elbow Room*. Cambridge, MA: MIT Press.

Dennett, D. C. 2003. *Freedom Evolves*. London: Penguin Books.

Gould, S. J., and R. C. Lewontin. 1979. The spandrels of San Marco and the Panglossian paradigm: A critique of the adaptationist programme. *Philosophical Transactions of the Royal Society B* 205:581–598.

Smith, A. 1976. *The Theory of Moral Sentiments*. Oxford: Oxford University Press.

Prologue: From the Beginning Was the Word

Evolutionary theory can be a nasty business. Perhaps it has something to say about human nature.

Most scientifically respectable evolutionary theories wear garments of math. I think of mathematical models as disciplined metaphors. We use x to represent something in the world, say slugs, and y to represent something else, say lettuces, then we analyze the relation of x to y using mathematics. We imagine that slugs and lettuces behave like x and y in the model, and then we use how x and y behave in the model to understand how slugs and lettuces behave in the world. Nobody can argue with mathematical models—that is one of the points of using mathematics—but there can be endless arguments about what you put into a model and what you leave out, and endless arguments about what the model means, because metaphors can be interpreted in many ways. I am not criticizing the use of metaphors—far from it, they are essential. All that we know about the world is metaphor. Our perceptions are a virtual reality, not the thing in itself but something that stands in the place of the thing. Phenomena are metaphors used to comprehend things. Don't worry, this book contains almost no mathematics;

but, if you don't like metaphor, then this is probably a good time to return the book and ask for a refund.

As an undergraduate, I was exhorted to eschew intentional idioms and, forty years later, I am still admonished for my "unscientific" language by anonymous reviewers who believe devoutly that such language should be suppressed. The moral tone in these criticisms suggests something important is at stake. Nobel Prize winners in the audience roll their eyes when I give public lectures. I am told that my choice of words expresses everything that is "soft" and unscientific about adaptationism as compared to the "hard" and rigorous sciences (*rigor* is stiffness in Latin). But when you listen closely to my critics, their own language abounds with reference to codes, signals, messages, and the like. When challenged, these tough-minded empiricists insist that these terms have strict physical meanings that do not invoke purposes. Most would reject, as absurd, a claim that they were using metaphor.

If it were only about words, I would have had better things to do than write this book, and you would probably have had better things to do than read it. I do not mean to disparage words. Spoken and written language are the expression of deep inner structures. The language that is censored says something about the values and fears of the censor. This book pays close attention to the meanings of words for four main reasons. The first is that languages evolve and provide useful analogies for thinking about genetic evolution. For example, the original reasons for particular turns of phrase can be forgotten, e.g., *exempli gratia*. The second is that meaning is the outcome of a process of interpretation and is specific to each interpreter. The same words will be interpreted differently and mean different things for each reader. As a result, many acrimonious disputes in the philosophy of biology are really quibbling about definitions rather than disputes about facts. The third is that the origin of language marked an extraordinary expansion in the lexical expressivity of the flux of meaning. The fourth, and most important, is that the beauty and diversity of language, like the beauty and diversity of the natural world, are wonders to behold.

Over the years I have come to believe that many promising approaches in biology, some of agronomic or medical significance, are not pursued because they violate the philosophical presumptions of working biologists who are unaware of their own presumptions. Good ideas are rejected for bad reasons. The money that goes into publicly funded research could be better spent. I see the denial of a naturalized teleology—a teleology grounded in the powerful metaphor of natural selection—as one of these self-defeating presumptions. It is a refusal to acknowledge the obvious, that organisms do things for good reasons.

The admonition that teleology has no place in science can be traced to the seventeenth century and the rejection of final causes as useful explanatory principles at the beginning of the Scientific Revolution. Final causes were one of the four categories of Aristotelian causes that had been a mainstay of medieval philosophy. In very rough terms, the *material cause* was the stuff out of which a thing was made; the *formal cause* was that which made of this stuff one kind of thing rather than another; the *efficient cause* was that which set a thing in motion; and the *final cause* was the purpose or end (*telos*), that for the sake of which the thing existed. The new materialist philosophy embraced material and efficient causes (matter in motion) but rejected final causes and had an ambivalent relation to formal causes. Lurking in the background was a separation of facts from values. An important scientific value was that facts were more important than values (irony intended), but if we were true scientists we should want to understand the origins of values and meanings.

The main task of this book is to explain how a physical world of matter in motion, of material and efficient causes, gave rise to a living world of purpose and meaning, of final and formal causes. I will briefly sketch the book's arguments so that you have some sense where it is heading. The key development in the history of life was the origin of writing, of materials that were copied and had effects in the world that directly or indirectly influenced a copy's chance of being copied. These *genetic* materials do not create new matter from the void but rearrange existing matter to match a model. The successive copies are material things, not incorporeal ideas, but lineages of material genes preserve structure despite perpetual change in molecular substance. What is "communicated" from model to copy? One may call it in*form*ation or simply *form*. Genetic materials can be considered formal causes. Aristotle would have said that the formal cause of our being human is that which makes of our material cause a human being. Human, chimpanzee, and slug bodies are built of the same materials but have different forms. Our bodies are more similar to the bodies of chimpanzees than to the bodies of slugs because we share more evolutionary history with chimpanzees than with slugs (one might say our formal causes are more similar).

But copying alone does not get us anywhere: garbage in, garbage out. We want to feed in garbage at one end and obtain something useful at the other end (an egg perhaps). Let's speak of eggs. Spinoza expressed a commonsense objection to final causes: the "doctrine of Final Causes turns Nature completely upside down, for it regards as an effect that which is in fact a cause, and vice versa. . . . It makes that which is by nature first to be last" (2002, 240). Common sense tells us that causes cannot come after their effects and a final cause appears to explain an earlier something by a later something, the goal the earlier something is intended to achieve. But common sense extrapolates an uncontroversial property of *causes of particular events* to a problematic restriction on *causes of types of events*. My favorite example, which will recur as a leitmotif throughout this volume, concerns chickens and eggs. It is unproblematic to ask whether this particular chicken was a cause of this particular egg. If the egg developed into the chicken, then the egg is a cause of the chicken. If the chicken laid the egg, then the chicken is a cause of the egg. But the question becomes problematic when we ask how generic chickens are causally related to generic eggs, because chickens then occur both before and after eggs. Eggs exist for the sake of becoming chickens and chickens for the sake of laying eggs.

Final causes emerge from replicative recursion by the process we call natural selection. Genes and their effects are like eggs and chickens. A gene's *effects* have a *causal* role in determining which genes are copied. A gene (considered as a lineage of material copies) persists if its lineage has been consistently associated with survival and reproduction. If possession of a gene is consistently associated with survival and reproduction then one can infer that the gene's *effects* have *causally* contributed to the gene's persistence. (A gene does not achieve anything except in the context of other genes and environmental inputs during a single life but, over the course of many generations, each gene is shuffled onto random backgrounds of other genes and this randomization allows natural selection to "infer" causation from correlation.) If past effects that have contributed to a gene's persistence are recapitulated as effects of a current gene, then these *effects exist for the sake of their cause* and can be considered the gene's raison d'être (or final cause). As a consequence of differential replication of alternatives, successful genes accumulate information about what has worked in the past. This information *comes from the environment* that selects.

From this perspective, natural selection can be viewed as inductive reasoning about effective action: a gene's effects are hypotheses about what works in the world, with confidence in a hypothesis increasing with the strength of past associations with favorable outcomes. But there is no guarantee that the world will not change. As David Hume wrote about inductive reasoning: "All inferences from experience suppose, as their foundation, that the future will resemble the past, and that similar powers will be conjoined with similar sensible qualities. If there be any suspicion that the course of nature may change, and that the past may be no rule for the future, all experience becomes useless, and can give rise to no inference or conclusion" (1748/2004, 22). What worked in the past need not work in the present because the world may have changed.

But surely, some will argue, evolutionary intentionality is ersatz intentionality, a metaphor, not the real thing. Only beings like us have real intentions because only we can foresee the future. But we cannot know the future. There is no backward causation. Our intentions are anticipated outcomes. When I act with intention, I have an outcome in mind. My deliberation is past and my plan may misfire. The unintended products of natural selection can likewise be considered agents whose bodies and instinctive behaviors "anticipate" that what has worked in the past will work in the future.

There remains a metaphysical sense in which all bodies of living things—including our own—can be considered nothing but matter in motion. Material causes of intricate structure have been shaped by unknowably complex chains of efficient causes that were set in motion before the origins of life. But this is a declaration of faith, not a practical procedure for understanding the world. Formal and final causes are human tools—furnaces and engines of discovery—that are both practical and fruitful for understanding living things. Their rejection as explanatory principles for biology throws the baby out with the placenta. The commandment of my undergraduate educators, *thou shalt not use teleological language*, is no more than a dogmatic insistence on ritual purity.

The intentionality of natural selection is retrospective. But *what worked* in the past was observing the present and predicting the future. Organisms have evolved to be real-time interpreters of their worlds, to be users of information *about* the world to guide effective action *in* the world. Organisms deliberate and decide. The latter chapters of this book define *meaning* as the output of a process of interpretation, but I cannot fully justify that claim here. It is what the book is *about*. I will argue that there is a continuum from the very simplest forms of interpretation, instantiated in single RNA molecules near the origins of life, to the type of interpretation that is going on in your mind as you skeptically evaluate this sentence. My hope for this book is that an appreciation of this continuum of meaningful interpretation will help to reunite the humanities and sciences in a continuum of intellectual endeavor. I hope you give me a chance to make my case.

So far I have talked about the book's contents and goals, but it would also help to say something about its form. This book is intended to mirror the products of natural selection, which themselves mirror the creative process. And this intention, like the products of natural selection, was partly retrospective. I began to compile a collection of previously published and unpublished papers that I needed to tweak at the margins so that they worked together as a whole but soon recognized that my editorial choices were a metaphor of the evolutionary process. When I first wrote the texts that comprise this book I could not know my final destination. Natural selection similarly has no purposes or preordained ends but creates beings with intended meanings. It acts to solve small local problems but, in the process of solving small problems, it inadvertently solves larger problems because everything it does is inadvertent. I have spent most of my life attempting to solve small local difficulties in evolutionary theory but hope in the process to have solved some larger problems.

Natural selection reuses old materials for new purposes. Its products are thereby comprised of parts of variable age that nevertheless must work together in some more-or-less coherent fashion. The resulting genomes are pastiche and so is this book. Its bricolage extends to the extensive use of quotations and paraphrase. If something has been said well, use it again. There is a saving of literary labor just as there is a saving of evolutionary labor. Darwin understood this principle well in a discussion of how old wheels, springs, and pulleys (slightly altered) can be repeatedly reused for new purposes, and I will repeatedly reuse his metaphor. One reason for not starting at the beginning and attempting to write a fully coherent whole that expresses how I see the world now is that I am a constantly changing fiction I tell myself. Some of my earlier selves may have understood the issues more clearly than I do now.

This book was not started at the beginning and written through to the end with the oldest bits first and the newest bits last. If you choose, you can read it the same way, skipping ahead and then coming back for a second reading. In some of the chapters, I descend into messy biochemical details rather than provide a highly simplified version (although my scientific readers will consider my presentation highly simplified). You may choose to scan these sections without trying to comprehend the details, because my principal intent is to show the underlying complexity of some of the simplest things we do. But if you get the deconstructionist bug for following the flux of meaning in biological bodies, you might give a close reading to how internal messages are continuously reinterpreted in new molecular media. Finally, I believe that the humanities and sciences have much to say to each other, so I wished to express my ideas in a style that would engage both audiences at the risk of enraging both and being ignored by both. Much of the prose was originally written under the constraints of meeting the selective criteria of scientific reviewers, and it shows. But the freedom from these constraints as I have revised the text has been liberating.

One of the unintended outcomes of thinking about what I meant by a gene was a new way of thinking about meaning. Meaning does not reside in the input to the reader but in the output: whatever the reader interprets a text to mean. I hope that some of your interpretations will be both complimentary and complementary, but once a text has been written its meanings reside with its readers. This book does not claim to have discovered a "truth" but to present ways of thinking about the world, and of interpreting words, that I have found useful and hope you might find useful.

Don't know much about history
Don't know much biology
Don't know much about a science book,
Don't know much about the French I took
But I do know that I love you,
And I know that if you love me, too,
What a wonderful world this would be.

—Sam Cooke, Lou Adler, Herb Alpert

1 Barren Virgins

> Like most philosophers of his age, he coquetted with those final
> causes which have been named barren virgins, but which might be
> more fitly termed the hetairæ of philosophy, so constantly have they
> led men astray.
>
> —T. H. Huxley (1869)

Francis Bacon failed to attain the preferment he desired under
Queen Elizabeth (1558–1603) but aspired to rise higher in the
administration of King James (1603–1625). His *Advancement of
Learning* (Bacon 1605/1885) was dedicated to the new king and
presented proposals for a reformation of education and scholarship with a shift in focus from book-learning and the classics to
practical arts and experimental science. Book II opened with a
direct address to James:

It might seem to have more convenience, though it comes often otherwise to pass (excellent king), that those which are fruitful in their generations, and have in themselves the foresight of immortality in their
descendants, should likewise be more careful of the good estate of future
times, unto which they know they must transmit and commend over
their dearest pledges. Queen Elizabeth was a sojourner in the world in
respect of her unmarried life, and was a blessing to her own times; and
yet so as the impression of her good government, besides her happy

memory, is not without some effect which doth survive her. But to your Majesty, whom God hath already blessed with so much royal issue, worthy to continue and represent you for ever, and whose useful and fruitful bed doth yet promise many the like renovations, it is proper and agreeable to be conversant not only in the transitory parts of good government, but in those acts also which are in their nature permanent and perpetual. Amongst the which (if affection do not transport me) there is not any more worthy than the further endowment of the world with sound and fruitful knowledge. (76–77)

Elizabeth was a hard act to follow, but Bacon flattered James by alluding to his greater sexual potency. The theme of fruitless females and fruitful males returned in a plea for funding experimental science:

Another defect I note, wherein I shall need some alchemist to help me, who call upon men to sell their books and to build furnaces, quitting and forsaking Minerva and the Muses as barren virgins, and relying upon Vulcan. . . . There will hardly be any main proficience in the disclosing of nature, except there be some allowance for expenses about experiments; whether they be experiments appertaining to Vulcanus or Dædalus, furnace or engine, or any other kind. And therefore as secretaries and spials of princes and states bring in bills for intelligence, so you must allow the spials and intelligencers of nature to bring in their bills; or else you shall be ill advertised. (80)

Bacon assigned physics and metaphysics to distinct domains of scholarship:

Physic should handle that which supposeth in nature only a being and moving; and metaphysic should handle that which supposeth further in nature a reason, understanding, and platform. . . . The one part which is physic, inquireth and handleth the material and efficient causes; and the other which is metaphysic, handleth the formal and final causes. Physic (taking it according to the derivation, and not according to our idiom for medicine) is situate in a middle term or distance between natural history and metaphysic. For natural history describeth the variety of things; physic the causes, but variable or respective causes; and metaphysic the fixed and constant causes. (114)

The focus of the new learning was to be empirical. Fixed and constant causes were of no practical use. Bacon rejected emphatically a role for final causes in the physical domain:

For the handling of final causes, mixed with the rest in physical inquiries, hath intercepted the severe and diligent inquiry of all real and physical causes, and given men the occasion to stay upon these satisfactory and specious causes, to the great arrest and prejudice of further discovery. . . . They are indeed but remoraes and hindrances to stay and slug the ship from further sailing; and have brought this to pass, that the search of the physical causes have been neglected and passed in silence. (118–119)

Bacon prospered under James, being appointed Solicitor General (1607), Attorney General (1613), Lord Chancellor (1618), Baron Verulam (1618), and Viscount St. Alban (1621). Then, he was found guilty of corruption by a parliamentary committee of his enemies and removed from public office (1621). He devoted his remaining years to scholarship. *De dignitate et augmentis scientiarum* was published in 1623 as an expanded Latin version of *The Advancement of Knowledge*. In a new section on the practical doctrine of nature, he compared final causes to fruitless females:

Physica siquidem et inquisitio causarum efficientium et materialium producit mechanicam: at metaphysica et inquisitio formarum producit magiam; nam causarum finalium inquisitio sterilis est, et, tanquam virgo Deo consecrata, nihil parit. (Bacon 1623/1829, 192)

In rough paraphrase: "Physics, the investigation of efficient and material causes, produces mechanics, whereas metaphysics, the investigation of forms, produces magic; but the investigation of final causes, like a virgin consecrated to God, gives birth to nothing." Final causes were without practical application. The principal referrents of *virgo Deo consecrata* for Bacon's contemporaries would have been Catholic nuns, but, by the nineteenth century, Bacon is commonly interpreted as having alluded to the vestal virgins and to have described final causes as "barren virgins." In my readings, Bacon never refers to final causes as "barren virgins," an epithet he reserved for Minerva and the Muses, although *virgo Deo consecrata nihil parit* meant more or less the same thing.

René Descartes's (1641/2011) Fourth Meditation similarly rejected a place for final causes in physics, in his case, because the mind of God was inscrutable:

Since I already know that my nature is very weak and limited and that the nature of God is immense, incomprehensible and infinite, I also know from this that there are innumerable things of which I do not know the causes. For this reason alone, I think there is no role in physics for that whole class of causes which are usually sought in purposes, because I think I cannot investigate God's purposes without temerity. (39)

Descartes returned to this theme in his *Principles of Philosophy* (1647/1983):

And so, finally, concerning natural things, we shall not undertake any reasonings from the end which God or nature set Himself in creating these things, and we shall entirely reject from our Philosophy the search for final causes: because we ought not to presume so much of ourselves as to think that we are confidants of His intentions. (14)

Neither Bacon nor Descartes rejected final causes outright, but both believed they should be kept in their place. And that place was not physical inquiry. The allocation of efficient and final causes to separate domains of inquiry has been a long-lasting compromise to which many would subscribe today. Thus William Whewell opined:

Final causes are to be excluded from physical inquiry; that is, we are not to assume that we know the objects of the Creator's design, and put this assumed purpose in the place of a physical cause. The physical philosopher . . . [makes] no use of the notion of final causes: and it is precisely because he has thus established theories independently of any assumption of an end, that the end, when after all it returns upon him and cannot be evaded, becomes an irresistible evidence of an intelligent legislator. . . . Bacon's comparison of final causes to the vestal virgins is one of those poignant sayings, so frequent in his writings, which it is not easy to forget. . . . If he had had occasion to develope his simile, full of latent meaning as his similes so often are, he would probably have said, that to these final causes barrenness was no reproach, seeing they ought to be, not the mothers but the daughters of our natural sciences; and that they were barren, not by imperfection of their nature but in order that they might be kept pure and undefiled, and so fit ministers in the temple of God. (1833, 266)

Form and Function

The seventeenth-century exclusion of formal and final causes from scientific explanation bore abundant fruit in investigations of the nonliving world, but nonphysical causes continued to be used as explanatory principles in the study of living beings: teleological concepts of the physiological (normal) and pathological (abnormal) were central to medicine and physiology; naturalists interpreted the intricate adaptations of plants and animals as evidence of a Creator; embryologists interpreted development as proceeding toward a final form. For theists, the compromise of separate explanatory domains of science and theology allowed bodies to be viewed as mechanisms whose form and function revealed the wisdom and beneficence of a divine legislator.

The structures of living things showed clear evidence of aptness for function, but organisms of different species exhibited structural similarities that were difficult to explain solely in terms of a shared purpose. Skeletons of sperm whales possessed pelvic bones, like other mammals, despite the absence of hind limbs. Ostriches had wings, like other birds, but were unable to fly. Mysterious laws of form appeared to exist independent of final causes. The living world contained individuals that belonged to species that could be grouped into genera that could be grouped into more inclusive categories on the basis of "affinities" revealed in correspondences of form or structure. No two individuals were exactly alike, but the members of a species belonged together as deviations from an ideal form. The "same" generic parts could be recognized in different species albeit with specific differences. Taxonomists attempted to bring order to the abundance of living forms with the search for a "natural" system of classification.

Morphology, a term coined by Goethe, developed in the nineteenth century as a science of form. Comparative anatomists found similarities of inner structure despite superficial differences of outward form and looked for nonmaterial principles of explanation. Georges Cuvier championed "des conditions d'existence, vulgairement nommé des causes finales" ("the conditions of existence, commonly named final causes") (1817, 6) as determining animal form, but others recognized aspects of form that were inexplicable in terms of function. Étienne Geoffroy Saint-Hilaire propounded a unity of organic composition behind the seductions of form and function. He envisaged a single plan expressed in the diverse forms of all animals (Le Guyader 2004). Their disagreements came to a head in a series of debates before the French Academy of Sciences in 1830. Cuvier was generally interpreted as having bested Geoffroy in these debates.

In *Indications of the Creator*, William Whewell expressed confidence in the doctrine of final causes:

The assumption of hypothetical final causes in physics may have been, as Bacon asserts it to have been, prejudicial to science; but the assumption of unknown final causes in physiology, has given rise to the science. The two branches of speculation, Physics and Physiology, were equally led, by every new phenomenon, to ask their question, "Why?" But, in the former case, "why" meant "through what cause?" in the latter, "for what end?" And though it may be possible to introduce into physiology the doctrine of efficient causes, such a step can never obliterate the obligations which the science owes to the pervading conception of a purpose contained in all organization. (1845, 21)

In distinctly Kantian terms, Whewell saw final causes as constitutive of understanding in both the organic and inorganic worlds but regulative of reasoning about the organic world:

This Idea of a Final Cause is applicable as a fundamental and regulative idea to our speculations concerning organized creatures only. That there is a purpose in many other parts of creation, we find abundant evidence to believe from the arrangements and laws which prevail around us. But this persuasion is not to be allowed to regulate and direct our reasonings with regard to inorganic matter, of which conception the relation of means and ends forms no essential part. In mere Physics, Final Causes as Bacon has observed, are not to be admitted as a principle of reasoning. But in the organical sciences, the assumption of design and purpose in every part of every whole, that is, the pervading idea of Final Cause, is the basis of sound reasoning and the source of true doctrine. (49)

Richard Owen had at first sided with Cuvier and final causes, but his further studies caused him to reconsider his position (Owen 1868, 787). Although he was not ready to derive all forms from a single transcendent plan, he nevertheless recognized a unity of type in the forelimbs of moles, horses, bats, and whales despite disparate functions (Owen 1849). Structural correspondences, in some ill-defined way, expressed "ideas" that were independent of use and details of form. As a result of these studies he found "the artifice of an archetype vertebrate animal was as essential as that of the archetype plant had been to Goethe in expressing analogous ideas; and as the like reference to an 'ideal type' must be to all who undertake to make intelligible the 'unity in variety' pervading any group of organisms" (Owen 1868, 788). He expressed the continuing discontent of many morphologists with final causes, using a familiar metaphor (Owen 1849, 40):

A final purpose is indeed readily perceived and admitted in regard to the multiplied points of ossification of the skull of the human foetus, and their relation to safe parturition. But when we find that the same ossific centres are established, and in similar order, in the skull of the embryo kangaroo, which is born when an inch in length, and in that of the callow bird that breaks the brittle egg, we feel the truth of Bacon's comparisons of "final causes" to the Vestal Virgins, and perceive that they would be barren and unproductive of the fruits we are labouring to attain, and would yield us no clue to the comprehension of that law of conformity of which we are in quest.

Darwinian Nuptials

By the mid-nineteenth century, the mechanical philosophy had secured the field of the physical sciences but the tangled bank of biology remained contested ground. Charles Darwin's notes from 1838 contain the following:

The Final Cause of innumerable eggs is explained by Malthus.—[is it anomaly in me to talk of Final causes: consider this!—] consider these barren Virgins. (Barrett et al. 1987, 637)

These cryptic words can be interpreted in many ways. An interpretation I find attractive is that "barren virgins" refers to the innumerable eggs (most will die without issue); the Malthusian struggle for existence explains why eggs are produced in such large numbers; and the parenthetical note is an injunction to consider whether natural selection eliminated final causes from the living world or explained them.

In *On the Origin of Species*, Darwin wrote:

It is generally acknowledged that all organic beings have been formed on two great laws—Unity of Type, and the Conditions of Existence. By unity of type is meant that fundamental agreement in structure, which we see in organic beings of the same class, and which is quite independent of their habits of life. On my theory, unity of type is explained by unity of descent. The expression of conditions of existence, so often insisted on by the illustrious Cuvier, is fully embraced by the principle of natural selection. . . . The law of the Conditions of Existence is the higher law; as it includes, through the inheritance of former adaptations, that of Unity of Type. (1859, 206)

Thus Darwin explained unity of type by transformation in *actual evolutionary time* rather than *abstract formal space* and was able to reconcile similarity of structure with divergence of function.

In one of the first reviews of *On the Origin of Species*, T. H. Huxley suggested that Darwin had charted a course past the sirens of teleology:

The path he bids us follow professes to be, not a mere airy track, fabricated of ideal cobwebs, but a solid and broad bridge of facts. If it be so, it will carry us safely over many a chasm in our knowledge, and lead us to a region free from the snares of those fascinating but barren Virgins, the Final Causes, against whom a high authority has so justly warned us. (1859, 9)

Huxley had missed the head of the nail. Darwin returned to the relation between past and current utility in *On the Various Contrivances by which British and Foreign Orchids Are Fertilised by Insects*:

Although an organ may not have been originally formed for some special purpose, if it now serves for this end we are justified in saying that it is specially contrived for it. On the same principle, if a man were to make a machine for some special purpose, but were to use old wheels, springs, and pulleys, only slightly altered, the whole machine, with all its parts, might be said to be specially contrived for that purpose. Thus throughout nature almost every part of each living being has probably served, in a slightly modified condition for diverse purposes, and has acted in the living machinery of many ancient and distinct specific forms. (1862, 348)

Darwin believed that investigation of the usefulness "of each trifling detail of structure is far from a barren search to those who believe in natural selection," but he added the following important caveat:

I do not here refer to the fundamental framework of the plant, such as the remnants of the fifteen primary organs arranged alternately in the five whorls; for nearly all of those who believe in the modification of organic beings will admit that their presence is due to inheritance from a remote parent-form. (1862, 352)

Huxley continued to grapple with the relation between natural selection and final causes. In a review of Ernst Haeckel's *Natürliche Schöpfungs-Geschichte*, he opined:

Perhaps the most remarkable service to the philosophy of Biology rendered by Mr. Darwin is the reconciliation of Teleology and Morphology, and the explanation of the facts of both which his views offer. (1869a, 13)

But this was a reconciliation without enthusiasm. Huxley now recognized two kinds of teleology. The first was

the Teleology which supposes that the eye, such as we see it in man or one of the higher *Vertebrata*, was made with the precise structure which it exhibits, for the purpose of enabling the animal which possesses it to see. (1869a, 14)

This kind of teleology had "undoubtedly received its death-blow" from the doctrine of evolution. The appearance of purpose could be explained by mechanical processes without recourse to final causes. Morphology had been reconciled with a corpse. The second was a "wider Teleology" that viewed the universe as a vast mechanism with an undisclosed purpose. This wider teleology was untouched by the doctrine of evolution but was a subject on which the scientific investigator must remain agnostic:

If the teleologist assert that this, that, or the other result of the working of any part of the mechanism of the universe is its purpose and final cause, the mechanist can always inquire how he knows that it is more than an unessential incident—the mere ticking of the clock, which he mistakes for its function. And there seems to be no reply to this inquiry, any more than to the further, not irrational, question, why trouble one's self about matters which are out of reach, when the working of the mechanism itself, which is of infinite practical importance, affords scope for all our energies? (1869a, 14)

Huxley still heeded Bacon's admonition that final causes were barren virgins irrelevant to practical inquiry.

Asa Gray announced that Darwin had not merely reconciled teleology and morphology but had united them in fertile marriage:

Let us recognise Darwin's great service to Natural Science in bringing back to it Teleology: so that, instead of Morphology *versus* Teleology, we shall have Morphology wedded to Teleology. In many, no doubt,

Evolutionary Teleology comes in such a questionable shape, as to seem shorn of all its goodness; but they will think better of it in time, when their ideas become adjusted, and they see what an impetus the new doctrines have given to investigation. They are much mistaken who suppose that Darwinism is only of speculative importance and perhaps transient interest. In its working application it has proved to be a new power, eminently practical and fruitful. (1874, 81)

Darwin wrote to Gray in rapid response:

What you say about Teleology pleases me especially, and I do not think anyone else has ever noticed the point. I have always said you were the man to hit the nail on the head. (F. Darwin 1898, 367)

What was "the point" that so gratified Darwin? Most commentators have interpreted this to be Gray's statement that Darwin had wedded Morphology and Teleology, but this reading has problems not least of which is Huxley's earlier statement that such a reconciliation was Darwin's "most remarkable service to the *philosophy* of Biology" (emphasis added). Perhaps, 'the point' was what Gray said about the *practice* of Biology and the impetus to investigation. Final causes were far from barren virgins. They were powers both *practical* and *fruitful*.

An Unconsummated Union?

In two groups of animal, however much they may at present differ from each other in structure and habits, if they pass through the same or similar embryonic stages, we may feel assured that they have both descended from the same or nearly similar parents, and are therefore in that degree closely related. Thus, community in embryonic structure reveals community of descent.

—Charles Darwin (1859)

In *On the Origin of Species*, Darwin argued that comparative embryology revealed resemblance among related forms because of common descent and differences because of divergent adaptation. He presented his theory of descent with modification as unifying what had been perceived as the conflicting demands of unity of type and conditions of existence. But Darwin's marriage of form and function was the unwitting source of an estrangement between the study of embryological development, a goal-directed process, and the study of evolutionary transformation by natural selection, a process without a preordained end.

For Aristotle, *telos* could refer to an endpoint toward which a thing moved or the utilitarian purpose that motivated an action. Since Aristotle, these senses of final cause, of goal and utility, had often been entangled. In the first half of the nineteenth century, "evolution" was generally used to refer to the process of development whereby a thing of one form (an egg or embryo) gave rise to a thing of another form (an adult). In this context, the end (*telos*) of the living thing was the achievement of its form (*eidos*). The use of "evolution" to refer to changes of form within a generation was sometimes extended to changes in form across many generations—the modern sense of the term—often with the associated belief that there was an analogy between the two processes of change and that understanding the former would provide insight into the latter. One might say that the emphasis of many nineteenth-century German morphologists was on the goal-directedness of development whereas the emphasis of British naturalists was on utilitarian function (German *Zweck* has stronger connotations of "goal" or "target" than English *purpose*).

From his extensive studies of the late-nineteenth century literature, the historian of science Peter Bowler (1983, 1992) concluded that *On the Origin of Species* catalyzed a widespread acceptance of evolutionary change among biologists but that Darwin's mechanism of natural selection was largely dismissed. Theories of evolutionary transformation were modeled instead on orderly, goal-directed processes of development. As one example, in *On Orthogenesis and the Impotence of Natural Selection in Species-Formation*, Theodor Eimer wrote:

Orthogenesis shows that organisms develop in definite directions without the least regard for utility through purely physiological causes as the result of *organic growth*. (1898, 2)

Such theories embraced the sense of final cause (directedness) that Darwin rejected but rebuffed the sense of final cause (utility) that Darwin embraced. The "evolutionary synthesis" or "modern synthesis" refers to an ill-defined period in the first half of the twentieth century during which evolution by natural selection was recognized as being compatible with Mendelian genetics. A common claim of those now calling for a reformation of evolutionary theory is that developmental biology was "excluded" from the synthesis of this earlier period. Another interpretation is that most leading embryologists in this period *chose* not to join because they were skeptical about the relevance of natural selection and Mendelian genetics for understanding processes of development (Hamburger 1980).

Reduction to Mechanism

Despite Darwin's attempted naturalization of organic purposes, the charms of evolutionary teleology were spurned by most experimental biologists. This was associated with a major shift toward explanations of living things in terms of physics and chemistry. During the nineteenth century, this shift can be seen first in physiology and then in the emerging fields of biochemistry, cell biology, and experimental embryology. The inorganic and organic worlds were seen as subject to the same fundamental laws. Because final causes had long been rejected by physicists and chemists, they were also rejected by those who wanted to subsume biology under chemistry and physics. The rise of mechanistic biology was well underway in the nineteenth century and became the dominant strain of twentieth-century biology.

As early as 1842, Emil du Bois-Reymond and Ernst Brücke had sworn to uphold the truth that the only forces active within organisms were physicochemical forces (du Bois-Reymond 1918, 108). Hermann Helmholtz and Carl Ludwig joined the oath-takers in a program to reform physiology in purely physicochemical terms (Cranefield 1957). Helmholtz spelled out the consequences of his principle of "conservation of force" for living things:

There may be other agents acting in the living body, than those agents which act in the inorganic world; but those forces, as far as they cause chemical or mechanical influences in the body, must be quite of the same character as inorganic forces, in this at the least, that their effects must be ruled by necessity, and must be always the same, when acting in the same conditions, and that there cannot exist any arbitrary choice in the direction of their actions. (1861, 357)

For most mechanists, the conservation of force negated the possibility that organisms could be unmoved movers, capable of arbitrary choices without prior physical cause.

The "mechanistic revolution" has been far more important than the "Darwinian revolution" in shaping the experimental practices and scientific philosophies of most modern biologists. Darwinism has not been central to their work. Some welcomed the justification that the evolutionary hypothesis provided for fundamental similarities in the inner workings of diverse organisms because it justified medically relevant research in yeast. Others saw Darwinism as a vindication of materialism and a justification for the elimination of final causes from nature. Many found the metaphor of natural selection to be scientifically suspect, especially its intentional overtones.

By the mid-twentieth century, mechanism had stormed the ramparts of biology and cleared the field for material and efficient causes. There were always some biologists who championed nonphysical causes, but they increasingly presented themselves as heretics against the rigid orthodoxy of hegemonic mechanism. During my own education, I was repeatedly warned against teleological thinking, and its close cousin anthropomorphism, by lecturers who spoke of the heart as a pump *for* the circulation of blood and of RNAs as messengers *for* the translation of proteins. The seventeenth-century rejection of a role for final causes in scientific inquiry has had long-lasting effects on the ideology of working scientists. Concepts of purpose and function shape the practice of biology—this is the natural way to think about living things—but the overt language of teleology is censored. Final causes are shunned as fruitless females who remain anathema to the virile vulcans of hard science.

2 Social Genes

The Selfish Gene (Dawkins 1976) was published the year before I entered Macquarie University as an undergraduate and remains in print more than forty years later with millions of copies sold. It is a book that arouses strong emotions. For many of its admirers, its lucid presentation of difficult ideas made new sense of life. For many of its critics, it was a bad (even evil) book. Some saw its message as "We are fundamentally selfish and our better natures are an illusion"; others objected to the denigration of organisms and glorification of genes; yet others objected to the superficiality of viewing cultural change as the selfish propagation of alternative memes. A common thread runs through many of these negative (and contestable) reactions, a skein of unacknowledged mind–body dualism. *The Selfish Gene* unabashedly used the language of agency and purpose in describing the products of natural selection. Many biologists, especially those seeking a union of the life and physical sciences, would expunge all talk of meaning and purpose from biology because this sullies the purity of their science. Others objected to Richard Dawkins using the language of agency for genes because agency was the proper preserve of human beings. Sometimes these critics were the same reader.

The Selfish Gene emphatically rejected popular notions that evolution acts for the "good of the group" or "good of the species." Natural selection, Dawkins argued instead, promotes the good of the individual or gene. *The Selfish Gene*, for the most part, equated individual and genetic fitness but favored the gene, rather than individual, as the fundamental beneficiary of adaptation. Genes were potentially immortal, passing through long series of mortal bodies, each of which was summarily discarded. These disposable bodies were viewed as elaborate survival machines evolved for the support and propagation of their genetic inhabitants.

"Good-of-the-group" arguments did not disappear, however. Group selection had its own passionate advocates who engaged in heated polemics with supporters of the selfish gene. David Sloan Wilson (1980), in particular, developed models in which individuals benefited from working together in groups. His mathematics was error-free, but the assumptions and interpretations of his models were vehemently attacked. His "groups," it was argued, were not really groups; or his models described individual-level selection, not group-level selection, because individuals benefited from being members of groups rather than sacrificed themselves for the group; and so on. Wilson responded in kind and hostilities escalated.

The defenders of groups argued that selection and adaptation can occur at more than one level: among groups of individuals, among individuals within groups, among cells within individuals, and among genes within cells. Selection of genes occurred at the lowest level of this hierarchy, but genes did not constitute a privileged level of fitness or adaptation. This perspective was generally labeled "hierarchical selection" in the 1980s but is now more commonly known as multilevel selection. The

Google Ngram Viewer suggests the crossover year, from numerical excess of "hierarchical selection" to a preponderance of "multilevel selection," was 1996. I suspect that political considerations contributed to this intriguing memetic shift, with the pluralistic and inclusive connotations of "multilevel" found more appealing than the authoritarian and aristocratic associations of "hierarchy."

It is impossible to read the arguments for and against multilevel selectionism without detecting political subtexts. The proponents of multiple levels of selection tended to present their models as showing that individuals (by implication ourselves) were kinder, gentler, and less selfish than how individuals were portrayed by gene-selectionists. This was resented by gene-selectionists who felt unfairly portrayed as defenders of selfishness. Beneath the acrimony, both camps were positing a less selfish individual, with selfishness either displaced onto genes or distributed across multiple levels. It seemed to me that the two sides were using a common vocabulary to mean different things. Their disagreements were largely semantic. Both sides were right in their own terms.

I did not read *The Selfish Gene* until sometime in the 1980s, and encountered the ideas of David Lack, George Williams, Bill Hamilton, Robert Trivers, and John Maynard Smith in their own words, rather than through Richard Dawkins's elegant prose. Therefore, *The Selfish Gene* did not change my life as many have told me it changed their lives, but I did read Dawkins's second book, *The Extended Phenotype*, soon after its publication in 1982, and that book was a revelation at the start of my doctoral studies. *The Extended Phenotype* was for me a radical book that undermined the unity of the individual. Dawkins wrote: "I see the concept of inclusive fitness as the instrument of a last-ditch rescue attempt, an attempt to save the individual organism as the level at which we think about natural selection." In Dawkins's view, a gene was subject to selection on how its *phenotype*—its effects in the world—promoted the gene's own propagation. These selectable effects could be *extended* beyond the boundaries of an individual's body into the inanimate environment and into the bodies of other individuals, even individuals of other species. Dawkins expressed his "central theorem" of the extended phenotype as follows: *"An animal's behaviour tends to maximize the survival of the genes 'for' that behaviour, whether or not those genes happen to be in the body of the particular animal performing it"* (1982, 233).

A second strand of *The Extended Phenotype* was dissension among genes within organisms. Different genes pursued divergent ends. Genes on the sex chromosomes favored different actions than genes on other chromosomes. Genes in the nucleus and mitochondria disagreed about the value of sons. Segregation distorters were outlaws that flourished at the expense of law-abiding Mendelian conformists. Transposable elements favored their own proliferation with scant regard for benefits to their bodily hosts. I became a convinced gene-selectionist and an advocate of intragenomic conflict. The remainder of this chapter, beginning with the next sentence, is the oldest stratum of the book you are currently reading.

The Extended Phenotype viewed the complex behaviors and structures that have evolved by processes of natural selection as adaptations for the good of the relevant genes (replicators) rather than for the good of individual organisms (vehicles). A common criticism of this view has been that organisms are integrated wholes, in which no gene can replicate without the assistance of many others. The implicit metaphor is of the organism as a machine, with genes as instructions for the assembly of the machine's component parts. But an alternative metaphor is possible: genes as members of social groups. Societies, like machines, can display intricate mutual dependence and elaborate divisions of labor; but, unlike machines, societies are not designed. Cooperation and coordination cannot be assumed; when present, they must be explained. Social theories vary in the causal relations they posit between individuals and the societies of which they are members: some theories emphasize the power of individual actions to shape society, whereas others emphasize the social constraints on individual freedom. This chapter views properties of organisms as social phenomena that arise from the actions of individual genes, and explores the internal conflicts that can disrupt genetic societies and the social contracts that have evolved to mitigate these conflicts.

Gene-centered theories are often reviled because of their perceived implications for human societies. But even though genes may cajole, deceive, cheat, swindle, or steal, all in pursuit of their own replication, this does not mean that persons must be similarly self-interested. Organisms are collective entities (like firms, communes, unions, charities, teams), and the behaviors and decisions of collective bodies need not mirror those of their individual members. As I write this paragraph, my replicators—my genes and my memes—are in constant debate, even dissension, yet somehow I muddle through. I am glad I am not a unit of selection.

Genes as Strategists

Genes are catalysts. They facilitate chemical reactions but are not themselves consumed. A gene influences its own probability of replication by the reactions it catalyzes, usually indirectly via transcripts and translated products. These effects can be likened to the gene's strategy in an evolutionary game. When, where, and in what quantity the gene is expressed is part of that gene's strategy to the extent that changes in the gene's sequence (mutations) could produce a different pattern of expression. The evolutionary theory of games (Maynard Smith 1982) has usually been phrased in terms of payoffs to individuals rather than their genes. This sleight of hand is possible because outcomes that enhance an individual's reproductive success also enhance the transmission of most (if not all) of the individual's genes.

Individual and genetic payoffs are no longer in close harmony when an individual's actions affect the reproductive success of relatives or when conflicts occur within an individual's genome. Interactions among relatives can be reconciled with an individualistic perspective by recourse to the concept of inclusive fitness (Hamilton 1964); but intragenomic conflicts pose a more intractable problem, because an individual's fitness, inclusive or otherwise, is ill-defined when different genes have different fitnesses. Such conceptual difficulties do not arise if genes, rather than individuals, are treated as the strategists.

Why use strategic thinking, which anthropomorphizes genes, instead of the well-developed infrastructure of population genetics? My reasons are pragmatic. Molecular biology reveals that genes are much more sophisticated than the stolid dominant or recessive caricatures of classical genetics. A gene may be expressed in some tissues, and some environments, but not in others; may have multiple alternative transcripts; may respond to signals from other genes; may have a history (be expressed when maternally derived but silent when paternally derived); and so on. Such complexities are difficult to model by traditional genetic methods. Game theory, however, allows evolutionarily stable strategies to be selected from among a large range of alternative patterns of gene expression. The realism of a strategic analysis depends on the realism of the set of alternatives from which candidate strategies are chosen. Some conceivable strategies may be unavailable in the real world, but too restricted a set of alternatives can also mislead.

Kinds of Genes

There are at least two defensible answers to the question "How many words in this book?" The first is the number tallied by the word processor of my computer. In this answer, a *word* is a string of characters terminated by a space or punctuation mark. Each time "vehicle" appears it is an extra word added to the tally. The second is the size of my vocabulary. In this answer, "replicator" counts as a single word no matter how many times it appears. *Gene* has a similar ambiguity. It can refer to the group of atoms that is organized into a particular DNA sequence—each time the double helix replicates, the gene is replaced by two new genes—or it can refer to the abstract sequence that remains the same gene no matter how many times the sequence is copied. The *material gene* (first sense) can be considered to be a vehicle of the *informational gene* (second sense). In philosophical parlance, the informational gene is a *type* of which material genes are the *tokens*.

Debates about the "units of selection" are interminable, partly because different meanings of "gene" are conflated. When hierarchical-selectionists describe the gene as the lowest level in a nested hierarchy of units (species, populations, individuals, cells, genes: Wilson and Sober 1994), their sense is closer to the material gene, whereas when gene-selectionists refer to the gene as the unit of selection (Dawkins 1982), their sense is closer to the informational gene. The informational gene may be represented materially at multiple levels of the vehicular hierarchy, but it is not itself a level of the hierarchy. On this view, the material gene is an ephemeral vehicle of the informational gene. However, the informational gene is not precisely the intended meaning of gene-selectionists. I will call their gene the *strategic gene* because their sense corresponds to the gene that is a strategist in an evolutionary game.

Every genetic novelty (new informational gene) originates as a modification of an existing gene and is initially restricted to a few vehicles at lower levels of the material hierarchy, solely because it is rare. Therefore, the gene's material copies will interact with each other only when they are present in different cells of the same body or in the bodies of closely related individuals. If such a gene is ever to become established, it must be able to increase in frequency under these circumstances. As the gene's frequency increases, its fate may be influenced by selection at higher levels of the hierarchy, but it will still retain the features that ensured its success when rare. Thus, the gene can be said to commit itself to a strategy when rare that it must maintain at all frequencies. The phenotypic effects of successful genes will consequently appear to be adaptations for the good of groups of material genes that interact because of recent common descent. A *strategic gene* corresponds to such a coterie of material genes and can be considered the unit of adaptive innovation.

The meanings of words (like genes) evolve, and it would be futile to legislate a single meaning of "gene," just as it would be futile to legislate a single meaning of "word." Semantic flexibility can even be useful when precise distinctions are unimportant, because it allows subtle shifts of sense without becoming embroiled in long terminological explanations. Occasional inconsistency is sometimes the price of brevity.

The Reach of the Strategic Gene

A strategic gene is defined by the nature of the interactions among the material copies of an informational gene that influence transmission of its sequence when its copies are rare. If all copies of the informational gene acted in isolation, the only phenotypic effects that would promote its transmission would be effects that directly promote the replication of its individual copies. A strategic gene would then be coextensive with a material gene. Material genes need not act in isolation. For example, material genes that are expressed in the soma of multicellular organisms do not leave direct descendants but promote the transmission of their replicas in the organism's germline. The strategic gene now corresponds to an organism-sized cluster of material genes. Similarly, a gene in the soma of one individual may promote the transmission of its copies in the germlines of relatives. In this case, the strategic gene becomes a cluster of material genes distributed among some, but not all, members of a family. Such a gene's strategy could be "Treat all offspring equally," not because all carry its copies, but because the gene has no way of directing benefits preferentially to the offspring who do.

If a gene copy confers a benefit B on another vehicle at cost C to its own vehicle, its costly action is strategically beneficial if $pB > C$, where p is the probability that a copy of the gene is present in the vehicle that benefits. Actions with substantial costs therefore require significant values of p. Two kinds of factors ensure high values of p: recognition (green beards) and relatedness (kinship). A "green beard" is present when genes are recognized directly or by their phenotypic effects. By contrast, genes recognize kinship by historical continuity: a mammalian mother learns to identify her own offspring in the act of giving birth; a male preferentially directs resources to the offspring of mothers with whom he has copulated; the other chicks in a nest are siblings; and so on. Under kin selection, a gene's strategy is blind to the outcome of each toss of the meiotic coin. Thus, the treatment of the members of a class of relatives does not depend on which genes they actually inherit, and p corresponds to a conventional coefficient of relatedness. By contrast, green beard effects will discriminate between brothers with and without the relevant gene.

Green beards gained their name from a thought experiment by Dawkins (1976), who considered the possibility of a gene that caused its possessors to develop a green beard and to be nice to other green-bearded individuals. Since then, a "green beard effect" has come to refer to forms of genetic self-recognition in which a gene in one individual directs benefits to other individuals that possess the gene. The recognition of self and the recognition of non-self are two sides of the same coin. Thus, the rejection of individuals that do not possess a label can also be considered a green beard effect, if the absence of the label is correlated with the absence of the genes responsible for rejection. Green beard effects have often been dismissed as implausible because a single gene has been considered unlikely to specify a label, the ability to recognize the label, and the response to the label. However, these functions could also be performed by two or more closely linked genes that are consistently inherited together.

The distinction between cooperation because of self-recognition (green beards) and historical continuity (kinship) is applicable to many genetic interactions. When homologous centromeres segregate at anaphase I of meiosis, their orderly behavior is made possible by the prior recognition of some degree of sequence identity between homologous chromosomes (a green beard), whereas when sister centromeres segregate at anaphase II, recognition is not necessary because the centromeres have been physically associated since their joint origin from an ancestral sequence (kinship). Similarly, the physical cohesion of the body is made possible because sister cells have remained in intimate contact since their origin from a common zygote (kinship), but this rich source of nutrients is defended against interlopers by an immune system that distinguishes self from non-self (a green beard effect).

The Prokaryotic Firm: Managing a Cytoplasmic Commons

Genetic replication makes use of energy and substrates that are supplied by the metabolic economy in much greater quantities than would be possible without a genetic division of labor. These resources are common goods, available to every gene in the cytoplasm. Thus, genetic communities are potentially vulnerable to free riders, genes that take more than they contribute, and the gains of trade from biochemical specialization would not have been possible without the evolution of institutions and of procedures that limit the opportunities for social exploitation. In particular, strict controls are expected on access to the machinery of replication.

DNA-based replicators are believed to have evolved from RNA-based replicators, possibly because DNA is copied with greater fidelity than RNA (Lazcano et al. 1988). The change also had implications for cellular security. Communities in which RNA polymerases were responsible for both replication and transcription would have been less easily policed than communities in which replication ("self-aggrandizement") was performed by DNA polymerases and transcription ("communal labor") by RNA polymerases. As a bonus, a community that modified its own genes to DNA, and periodically cleansed its cytoplasm with ribonucleases, would in the process eliminate most RNA-based parasites.

An effective way to manage the cytoplasmic commons is to link genes to a single origin of replication, and to exclude non-members from the cytoplasm. The chromosome becomes a team whose members' interests coincide. The solution is egalitarian, at least within the group. Each gene that joins the chromosome has equity and replicates once per cycle, no matter what its contribution during that cycle. Efficiency might conceivably be improved if genes that contributed more to productivity in the local environment were rewarded with increased copy number, but this argument ignores the costs of negotiating fair shares and of policing complex rules. The suppression of internal conflict by replication from a single origin has a price because genes can be copied more quickly from multiple origins than from one (Maynard Smith and Szathmáry 1993).

Dangerous Liaisons

Leda Cosmides and John Tooby (1981) called a set of genes that replicated together, and whose fitness was maximized in the same way, a *coreplicon*. They argued that intragenomic conflicts are likely if an organism contains more than one coreplicon, because the members of a coreplicon will sometimes be selected to maximize their own propagation in ways that interfere with the propagation of other coreplicons. Selection for short-term replication and selection for long-term replication may be opposed. A coreplicon that replicated faster than other coreplicons within its cell lineage would increase in frequency. However, if differential replication were costly for cell survival, the cell lineage would eventually be eliminated in competition with other lineages. Thus, the long-term interests of coreplicons that share a cell lineage will coincide if they never have opportunities to form new combinations with other genes in other lineages. *Recombination* decouples genes' fates and is therefore essential for the indefinite persistence of intragenomic conflict (Hickey 1982).

Many bacteria contain multiple circular genomes. By convention, one of these circles is designated the bacterial chromosome, and the extra circles are called plasmids. Plasmid replication consumes energy and substrates. Whether a plasmid pays for its keep—from the perspective of chromosomal genes—depends on the metabolic skills that its genes bring to the cell, on whether these skills are required in the current environment, and on the degree of coadaptation between plasmid and chromosome. Many of the genes for antibiotic resistance that are the scourge of modern hospitals are carried on plasmids. Such a plasmid may be essential in the presence of antibiotics but a burden in their absence (Eberhard 1980).

Most plasmids promote conjugation between their host and other bacteria, although some smaller plasmids rely on larger plasmids for these functions. In the process, a copy of the plasmid is retained by the donor cell and an uninfected bacterium acquires a plasmid. Thus, conjugation allows plasmids to colonize new cytoplasms. Chromosomal genes, by contrast, are usually not transferred. Therefore, the chromosome bears the costs of replicating the donated plasmid, and the costs of increased exposure to viruses during conjugation, but the donor chromosome seemingly gains little in return. If the plasmid encodes beneficial functions, these are transferred to a potential competitor. Perhaps, if a plasmid is a burden to chromosomal genes, the chromosome benefits from sharing its cold with rivals.

Plasmids cannot be categorized simply as parasites or mutualists. For example, a plasmid that initially reduced its host's competitiveness enhanced fitness after plasmid and chromosome were propagated together for five hundred generations (Lenski, Simpson, and Nguyen 1994). A plasmid has two modes of transmission, vertical and horizontal, and selection can favor its propagation by either path. Selection for greater horizontal transmission, at the expense of vertical transmission, will generally increase the costs of the plasmid to chromosomal genes, whereas selection for increased vertical transmission will generally benefit the chromosome. In the limit, when there is no horizontal transmission of the plasmid, the long-term fates of chromosome and plasmid are inexorably linked and they effectively become a single coreplicon. Similar arguments apply to the viruses, transposons, and other coreplicons that populate bacterial cytoplasms.

Protection Rackets

Plasmids, once acquired, are difficult to discard. Plasmid genes encode multiple functions that ensure their stable transmission within an infected lineage (Nordström and Austin 1989). Many plasmids encode a persistent "poison" and its short-lived "antidote." Thus, if a cell segregates without the plasmid, it is cut off from its supply of antidote and succumbs to the poison. The gene for the poison can be said to recognize the presence or absence of the gene for the antidote. Because poison and antidote are inherited as a unit, the plasmid can be said to recognize itself (a green beard effect). Such protection rackets take many forms (Lehnherr et al. 1993; Salmon et al. 1994; Thisted et al. 1994). Some plasmids, for example, encode a methylase and its matching restriction enzyme. The methylase modifies bacterial DNA so that it is protected from the restriction enzyme, which cleaves unmodified DNA. Bacteria that lose the plasmid die, because methylation must be restored each time the chromosome replicates. The restriction enzyme simultaneously defends its cytoplasm against viruses and rival plasmids that lack the appropriate methylase, just as a gangster defends his turf (Kusano et al. 1995).

Mitochondria of trypanosomes contain a large DNA maxicircle and many small minicircles. The maxicircle encodes essential genes in garbled form, whereas the minicircles encode guide RNAs that edit the otherwise unreadable transcripts to yield translatable mRNAs (Benne 1994). Could RNA editing have evolved as a minicircle-maintenance system? If so, one would predict that minicircles could also edit DNA and encrypt maxicircle genes in ways that only they could decipher. Minicircles spread from one mitochondrial lineage to another (Gibson and Garside 1990), strengthening their analogy with bacterial plasmids.

Team Substitutions

A nonrecombining bacterial chromosome is a team that does not change its members (except by mutation). Its social contract is "All for one, and one for all," not "Every gene for itself." Chromosomal recombination occurs on rare occasions as a coincidental side effect of the horizontal transfer of plasmids (conjugation) or viruses (transduction). Some bacteria, however, have evolved mechanisms by which DNA is taken from the environment and used to replace homologous sequences of the chromosome, so-called *natural transformation*. Transformation, unlike conjugation and transduction, is controlled by chromosomal genes (Stewart and Carlson 1986). Uptake of DNA is induced under conditions of nutritional stress and may have evolved primarily as a means of gaining nutrients (Redfield 1993). Nevertheless, the expression of DNA-binding proteins that prevent the degradation of the donor sequence and the induction of the enzymatic machinery of recombination suggest that recombination is not a mere side effect but has been positively selected (Lorenz and Wackerknagel 1994; Stewart and Carlson 1986).

Why should a team replace one of its members? The repair hypothesis views transformation as a means of replacing injured team members (damaged DNA). However, repair is unlikely to be the principal function of transformation because uptake of DNA is not induced by damage to the chromosome (Redfield 1993). The recombinant-progeny hypothesis views transformation as a means of trying out new players. Replacement of one gene by another occurs in a single cell of a clone and does not commit other team members to the new combination, because the old combination survives in other cells of the clone. A team's chances of remaining successful in a changing environment will presumably be improved by some degree of experimentation with new combinations. The problem is that, for each member of the team, the advantages arise only from changes at positions other than its own. A gene benefits from replacing an established gene on another chromosome, but suffers from being so replaced. The important social question becomes whether some positions are privileged and not subject to replacement—in particular, whether the genes responsible for transformation are themselves transformed.

Multicellular Corporations

The development of resistant spores by *Bacillus subtilis* illustrates the differentiation of soma and germline in simple form. A bacterium undergoes an unequal cell division to produce a mother cell (soma) and a prespore (germline). The mother cell engulfs the prespore, assists in formation of the spore coat, and is then discarded. The process is coordinated by an exchange of signals between mother cell and prespore (Errington 1996). The genes of the mother cell sacrifice themselves for their replicas in the spore. Some bacterial somas are more complex. *Myxococcus xanthum* is a motile, predatory bacterium that forms a multicellular fruiting body. Individual myxobacteria forage and divide in the soil, but, when nutrients become scarce, they aggregate to form a structure in which sacrificial stalk cells (soma) raise myxospores (germline) above the substrate (Shimkets 1990).

Organisms develop somas to gain the benefits of a cellular division of labor. A soma, however, is a rich resource that can be exploited by genes of other germlines. Therefore, the advantages of somatic specialization can be realized only if the genes of the soma have some degree of confidence that their copies are represented among the beneficiaries of their labor. The simplest means by which genes in somatic cells ensure that their efforts are well directed is physical cohesion between soma and germline. The genes of the *Bacillus* mother cell can be assured that their copies are present in the prespore because cell division and sporulation take place within an enclosed sporangium that excludes outsiders. However, as somas become larger and more complex, interactions between somatic cells and germ cells become less direct. This creates additional opportunities for parasites to misappropriate somatic effort, and necessitates more elaborate security systems to protect the soma from exploitation. The genes of my liver are almost certain to have copies in my testes (because of my body's physical cohesion) but my lump of lard and mass of meat would not last long without a sophisticated system of immune surveillance. This example shows how relatedness (cohesion) and green beard effects (immune surveillance) interact to maintain somatic cooperation.

Uninterrupted physical cohesion cannot protect the genes of somatic cells from exploitation if sister cells become detached to forage (as in *Myxococcus*) or to form complex organs (as in multicellular animals). Some form of cellular recognition is required. When two cells meet, their responses can be influenced by what they learn about each other. Molecules on their surface can provide clues about which genes are present in a cell and whether the cell is friend, foe, or indifferent. Two categories of molecular interactions can be distinguished. Homotypic interactions occur between identical molecules on the two cells and are a particularly direct means for a gene to recognize itself in other cells. Heterotypic interactions occur between molecules encoded by different genes and can also provide a gene with information about its presence or absence in another cell if there is linkage disequilibrium between the interacting genes (Haig 1996a). Thus, green beard effects may play an important role in the somatic security systems of multicellular organisms (particularly making use of the complete linkage disequilibrium between genes of different species).

The origin of molecules that were able to discriminate between themselves and closely similar molecules greatly expanded the strategies available to genes and made possible the evolution of large multicellular bodies. The ancestor of the immunoglobulin superfamily probably interacted with itself in homotypic adhesion or signaling, but the family now includes many heterophilic adhesion molecules as well as the T-cell receptors, MHC antigens, and immunoglobulins of the vertebrate immune system (A. F. Williams and Barclay 1988). The cadherins, to take another example, are a family of cell-surface proteins that bind to copies of themselves on other cells. Nose, Nagafuchi, and Takeichi (1988) introduced the genes for P-cadherin and E-cadherin into a cell line that lacked cadherin activity, creating two sublines that were identical except for this single gene difference. When the cells were mixed, they spontaneously segregated into discrete populations, like oil mixed with water. Cadherins play a pivotal role in organogenesis, but similar mechanisms could clearly be used to distinguish self from non-self.

A Chimeric Menagerie

Slime molds are eukaryotes with a life cycle remarkably similar to *Myxococcus* (Kaiser 1986). They feed as unicellular amoebae, but aggregate when starved to form a fruiting body with a simple division of labor between spores and somatic stalk. Slime molds are thus particularly vulnerable to somatic exploitation because there is no guarantee that the amoebae who respond to an aggregation signal are members of the same clone, or that a predator will not use the signal to lure amoebae to their doom. The dangers are real, although somewhat mitigated by the mechanisms of cell-surface recognition discussed in the previous section. *Dictyostelium caveatum* is a predator that responds to the aggregation signals of other species and devours their amoebae before forming its own fruiting body (Waddell 1982). Zygotes of *Dictyostelium discoideum* produce the aggregation signal and devour haploid amoebae of their own species as they respond to the signal (O'Day 1979). Some strains of *Dictyostelium discoideum* form chimeric fruiting bodies with amoebae of other strains without contributing to the stalk (Buss 1982).

Chimerism between members of a single species has also been described in animals. Vascular fusion frequently occurs between neighboring genotypes of the colonial urochordate *Botryllus schlosseri*. The progenitors of germ cells circulate in the blood and will colonize, and in some cases totally replace, the gonads of the neighboring soma (Pancer, Gershon, and Rinkevich 1995). As another example, "hermaphroditic" females of the haplodiploid scale insect *Icerya purchasi* are host to spermatogenic cells derived from sperm that entered the cytoplasm of an egg, but which failed to fertilize the egg nucleus because they were preempted by another sperm (Royer 1975). Thus, a sperm that fails to fertilize the eggs of the mother can try again with those of the daughter or granddaughter, or persist as a permanent haploid inhabitant of female somas. Occasional winged males are produced from unfertilized eggs (Hughes-Schrader 1948) but must compete for fertilization with the "reduced males" resident in female gonads.

Marmosets and tamarins regularly produce dizygotic twins in a uterus ancestrally designed for singletons (that is, in a *simplex* uterus, which lacks long uterine horns to keep squabbling offspring apart). The placental circulations of the twins fuse with the result that each adult marmoset carries blood cells derived from its twin (Benirschke, Anderson, and Brownhill 1962). If germ cells were also transferred, and equally mixed between twin brothers, the genes of their respective somas would be indifferent about which brother copulated, although competition within the brothers' testes and ejaculates could be intense. Chimerism between dizygotic twins is the rule for marmosets but is the exception for human twins (van Dijk, Boomsma, and de Man 1996). It is common, however, between human mothers and their offspring. Fetal cells circulate in a mother's blood from the early weeks of pregnancy and descendants of these cells may persist in a mother's body for decades after the child's birth (Bianchi et al. 1996). Are these cells simply lost, or do they manipulate the maternal soma for the offspring's benefit?

Exploitation of host somas by pathogens and parasites remains a major problem for multicellular organisms. This section's collection of intraspecific chimeras emphasizes that the risk of somatic exploitation is not restricted to members of different species. Of course, somatic exploitation within species usually involves the everyday strategies of coercion and deceit.

The Nuclear Citadel

The speed of replication limits the amount of DNA that can be efficiently copied from a single origin of replication. The chromosome of *Escherichia coli* takes about forty minutes to replicate (Zyskind and Smith 1992). If the thousand-fold larger genome of *Homo sapiens* were similarly organized as a circular chromosome with a single bidirectional origin of replication, it would take almost a month to replicate (Fonstein and Haselkorn 1995;

Morton 1991). Humans and other eukaryotes avoid this problem by using multiple origins of replication. The attendant risk that some parts of the genome will replicate faster than others is exacerbated because the alternation of gametic fusion and meiotic segregation creates ample opportunities for rogue elements to colonize new genomes (Hickey 1982). For these reasons, eukaryotes are expected to have evolved sophisticated systems for controlling unauthorized replication.

Two characteristic features of eukaryotic cells probably contribute to replicative security. The first is the separation of the machinery of protein synthesis (in the cytoplasm) from the genetic material (in the nucleus). Passage of large molecules to and from the nucleus is controlled at the nuclear pore complex. Before a protein can dock with this complex, it must possess nuclear localization signals that are recognized by docking molecules in the cytoplasm (Davis 1995; Hicks and Raikhel 1995). The second is the eukaryotic cell cycle. Replication is confined to a specific S phase. Before DNA can replicate, it must acquire a "replication licensing factor" that authorizes it to replicate once, but once only, per cycle (Rowley, Dowell, and Diffley 1994; Su, Follette, and O'Farrell 1995). The origin recognition complex (ORC) that marks a site for future initiation of replication causes transcription to be silenced in its vicinity (Rivier and Pillus 1994). Thus, genes near an ORC are prevented from producing locally acting RNAs that could tamper with the genes' own replication.

The bread mold *Neurospora crassa* has evolved a highly effective defense against genetic elements that replicate more than once in a cell cycle. If a sequence is repeated within a haploid nucleus, both copies are inactivated by methylation and subject to a process of repeat-induced point mutation (RIP) until their sequences have diverged sufficiently to be no longer recognized as similar (Selker 1990). Thus, if a DNA sequence replicates faster than other members of its collective, both the additional copies *and* the master sequence are corrupted by a process of programmed mutation.

Vertebrates compartmentalize their DNA into active regions and methylated regions that are maintained in a compact transcriptionally inactive state (Bestor 1990; Bird 1993). The inactive portion of the genome often contains large amounts of simple repetitive sequences that do not encode proteins and which are subject to high rates of sequence turnover because of replication slippage and unequal crossing over (Dover 1993). This arrangement may function, in part, as a system of defenses against intragenomic parasites. First, a higher proportion of insertions will occur in noncritical sequences. Second, foreign DNA (once inserted) is transcriptionally inactivated by a methylation process that may specifically recognize structural features of parasitic DNA (Bestor and Tycko 1996). Third, inserted DNA is subject to sequence degradation by replication slippage.

It would be misleading to argue that the sole function of the organizational changes of the eukaryotic nucleus has been internal security. Even when agents have identical interests there is still a problem of coordination. The genome of *E. coli* contains about 4,000 genes whereas the genomes of humans, mice, and pufferfish contain perhaps 20,000 different protein-coding genes. Bird (1995) has argued that the nuclear envelope and histone proteins of eukaryotic cells, and the extensive methylation of vertebrate chromosomes, are adaptations for the reduction of the transcriptional noise associated with larger genomes. New mechanisms of control and security measures would have evolved hand in hand.

The Sexual Revolution

Bacterial recombination involves the formation and dissolution of partnerships between coreplicons or the substitution of one gene for another in a process that has clear winners and losers. By contrast, meiotic recombination involves a symmetric relationship in which two temporary teams come together, swap members, and form new temporary teams. The members of successful teams get to play more often in the next generation than members of unsuccessful teams. Thus, a successful player is one who performs well as a member of many different teams, and the system favors teams of champions rather than champion teams. Team members pursue the same goals, not because their long-term destinies are indissolubly linked, but because the rules of meiosis ensure that all receive the same opportunities if only their team can make it through to the next lottery.

If every gene assorted independently at meiosis, players could not form long-term partnerships because any two players present in a haploid team before gametic fusion would have an even chance of parting at meiosis. This 50 percent probability of recombination per generation applies to almost all randomly chosen pairs of genes in organisms with multiple chromosomes. Genes that are linked on the same chromosome can expect to remain associated for longer periods. If some combinations of linked genes work more effectively together than others, these combinations will tend to occur in successful teams and leave more descendants than less favored combinations. By this process, selection generates nonrandom associations of players (i.e., linkage disequilibrium), but these associations are constantly being disrupted by recombination.

One of the major preoccupations of evolutionary genetics has been the question why so many genetic collectives regularly break up successful teams to take a chance on untried combinations. Zhivotovsky, Feldman, and Christiansen (1994) summarized numerous models that reached the conclusion "that, in a random mating population, if a pair of loci is under constant viability selection (the same in both sexes), with recombination between them controlled by a modifying gene, and if this system attains an equilibrium at which the major genes are in linkage disequilibrium, then new alleles at the modifying locus can invade only if they reduce the rate of recombination between the major loci." A similar principle applies for an arbitrary number of loci (Zhivotovsky, Feldman, and Christiansen 1994). The intuitive explanation of this "reduction principle" is that new teams generated by recombination will, on average, be less successful than existing teams that have survived a generation of selection. Therefore, individual players are more likely to be successful in the next generation if there is less recombination of their current team.

Despite the reduction principle, recombination is widespread in nature. One or more of the assumptions of models that predict selection for reduced recombination must be violated. Genes that increase recombination can be favored if a population has not reached selective equilibrium, because recombination increases the efficiency with which currently favored players are brought together in the same team. Technically, the advantage a team gains from having both gene A and gene B must be less than the sum of their individual contributions to team success (Barton 1995). A similar process favors increased recombination if the cost of injury (mutation) to A and B is greater than the sum of the costs if A and B were damaged individually (Charlesworth 1990). In both these examples, increased recombination improves the efficiency of selection because it reduces the risk that inferior players will "hitch-hike" on the success of their teammates, or, what amounts to the same thing, that superior players will be dragged down by lesser players. Theories that ascribe the adaptive advantage of recombination to increased resistance to parasites are explanations of this sort because recombination rates evolve in a constantly deteriorating environment in which the most-favored allelic combination is always in flux (Hamilton, Axelrod, and Tenese 1990).

The Open Society and Its Enemies

The reduction principle also breaks down in the presence of multilocus green beard effects. Green beard effects allow genes to direct benefits to teams in which they have a high probability of being present. As we have seen, a gene (or coalition of genes) can profit from conferring a benefit B on another team at cost C to its own team, if $pB > C$ where p is the probability that the gene (or coalition) is present in the team that benefits. If this probability were the same for all genes in the donor team, all members would gain equally from the transaction. But, if the probabilities differ—as they do when benefits are directed to green-bearded relatives at the expense of other relatives (Ridley and Grafen 1981)—some team members will lose while others gain. Linkage disequilibrium can enable small coalitions of genes to conspire against the common good, but high levels of recombination will disrupt the persistent nonrandom associations on which multilocus green beard effects depend. Other team members would suspect the motivation of a group of players who were simultaneously members of a rival team, and can benefit from disrupting cliques before they form.

The best-studied conspiracies are systems of meiotic drive. A haplotype in a heterozygous diploid causes the failure of gametes that do not carry its copies, usually by means of a two-locus poison–antidote mechanism. If the haplotype does not go to fixation, it must be associated with countervailing fitness costs that will be experienced in full by team members that are unlinked to the haplotype. Therefore, selection at unlinked loci favors increased recombination to disrupt the conspiracy and separate the poison from its antidote (Haig and Grafen 1991). Leigh (1971; see also Eshel 1985) has compared the genome to "a parliament of genes: each acts in its own self-interest, but if its acts hurt the others, they will combine together to suppress it." Segregation distortion and related phenomena are departures from fairness. "The transmission rules of meiosis," Leigh suggested, "evolve as increasingly inviolable rules of fair play, a constitution designed to protect the parliament against the harmful acts of one or a few. . . . Just as too small a parliament may be perverted by the cabals of a few, a species with only one, tightly linked chromosome is an easy prey to distorters."

The Eukaryotic Alliance

Most internal conflicts within the nucleus are defused by the procedures of fair segregation and allelic recombination. However, eukaryotes also contain genes, in mitochondria and plastids, that are not part of the meiotic compact. The eukaryotic cell originated as an alliance between nuclear genes and the genes of symbiotic bacteria. Many of the latter eventually joined the nuclear firm, but some retained a limited independence as the

mitochondrial and plastid genes of today. We do not fully understand why some genes have accepted (or been granted) nuclear
equity whereas others have maintained a separate contractual
arrangement, nor why these shifts of allegiance have been predominantly one-way, from organelle to nucleus. Nuclear and
organellar genes are mutually dependent, yet their different rules
of transmission can be a source of conflict in their partnership.

If different organellar lineages occupied the same cytoplasm
after gametic fusion, the lineages would be expected to compete for occupation of the cytoplasm, with concomitant costs
to nuclear genes. Cosmides and Tooby (1981) suggested that
nuclear genes have been selected to minimize conflicts among
organellar genes by causing the destruction of the organelles of
one gamete, either before or after the fusion of gametes. For this
reason, they proposed, the nuclear genes of one kind of gamete
(sperm) discard their organellar partners before fertilizing a different kind of gamete (eggs) that retains its organelles (for related
arguments, see Hastings 1992; Hurst and Hamilton 1992; Law
and Hutson 1992). Nuclear-enforced suppression of cytoplasmic
conflict may thus have been the key factor in the evolution of
eggs and sperm, with all other differences between the sexes arising from this initial dichotomy. In support of this conjecture,
Hurst and Hamilton (1992) have noted that morphologically
distinct sexes are absent in taxa that exchange nuclear genes
without cytoplasmic fusion.

Uniparental inheritance of mitochondria and plastids resolves
one conflict but creates another. Nuclear genes are transmitted
by sperm and eggs, whereas organellar genes are transmitted
by eggs alone. Organellar genes would therefore benefit from
preventing reproduction by male function, if this increased
the resources available for female function. Cytoplasmic male

sterility has evolved many times in flowering plants. In all well-studied cases, male sterility is caused by mitochondrial genes, but their effects are often countered by nuclear genes that restore male fertility. Chloroplasts also have predominantly maternal inheritance but chloroplast genes are not known to cause male sterility (Saumitou-Laprade, Cuguen, and Vernet 1994). The plastid genome may lack mechanisms to abort male function, or, if such mechanisms exist, they may be easily circumvented by nuclear genes.

Despite its internal conflicts, the eukaryotic alliance has been an outstanding success. Daniel Dennett (1995, 340–341) considers human beings to be a radically new kind of entity, comparable in importance to the eukaryotic cell. In his view, we are a symbiosis between genetic replicators and cultural replicators (memes). Just as eukaryote cells cannot survive without both nucleus and organelles, we cannot survive without both genes and memes; neither genes nor memes are dispensable; and neither genes nor memes can claim priority as representing our true selves. Genes and memes have very different rules of transmission, and a meme cannot simply be incorporated into a chromosome where it follows the rules of meiosis. Conflicts are therefore expected. Some people will die for an idea. Others will abandon their faith for a sexual fling.

Sex Chromosomes

An average gene from a species with two sexes spends equal time in male and female bodies because every individual has a mother and a father. However, at a selective equilibrium some genes (or combinations of genes) may be more successful than average in one sex and less successful than average in the other. Such sexually antagonistic genes will benefit from being associated with other genes that bias sex determination toward the sex in which they have a relative advantage (Rice 1987), whether this is a conventional fitness advantage (viability) or a segregational advantage (meiotic drive). The process is self-reinforcing because genes that influence sex determination spend more time in one sex than the other and can thus persist in linkage disequilibrium with genes whose disadvantage in the less frequent sex is greater than their advantage in the more frequent sex. Genomes thus have a tendency to split into factions that spend equal time in the two sexes (autosomes) and factions that specialize in one sex or the other (sex chromosomes).

Meiotic drive in spermatogenesis or oogenesis (but not both) can favor the evolution of sex chromosomes, because a distorter has an advantage in one sex that is absent in the other. Segregation distorters will also be favored if they arise on existing sex chromosomes. Associations between agents of meiotic drive and the genetic determiners of sex result in biased sex ratios, but these biases will be opposed by the parliament of genes (or at least by its autosomal majority). Half of the genes in the next generation will come from males and half from females. This means that members of a minority sex will leave more descendants on average than members of the majority sex, and autosomal genes will benefit from being present in the minority sex (assuming that the sexes are equally costly). Autosomal genes are expected to enforce fair segregation of sex chromosomes in the heterogametic sex because neither sex will then be in a majority.

Hamilton (1967) recognized that autosomal genes will sometimes favor biased sex ratios. He considered a model in which small numbers of unrelated females founded local populations, their offspring mated among themselves, and the newly mated females dispersed to found new local populations. If males were heterogametic and segregation were strictly Mendelian, the sex ratio in the global population would be very close to unity, with some variation among local populations because of random fertilization by X- and Y-bearing sperm. The expected fitness of a female offspring would be the same as the average female fitness for the global population, regardless of the local sex ratio, but the expected fitness of a male offspring would increase with the local proportion of females. Therefore, an autosomal gene that caused itself to be present in female-biased local populations would have higher than average fitness, and a balanced sex ratio would no longer be the unbeatable strategy. In this example, the parliament of genes contains a number of parties with different policies concerning the sex ratio. The X party, the autosomal party, and the mitochondrial party would enter into coalition against the Y party to force a female-biased sex ratio among offspring, but the coalition partners would lack unanimity about precisely which ratio (Hamilton 1979). Sexual politics can profoundly destabilize the "parliamentary rules" of meiosis (Haig 1993a).

Genomic Imprinting and the Altercation of Generations

Relatives are genetic collectives that share some, but not all, of their members. A gene can benefit from employing a contingent strategy that treats collectives differently depending on information about the probability r that a collective includes one of its copies. This section will assume that a gene's only information about r comes from the family tree (pedigree of collectives) and the Mendelian probabilities associated with the pedigree, in some cases supplemented with knowledge of parental origin and the uncertainty of paternity. Green beard effects will not be considered.

The simplest relationship between a diploid mother and her sexually produced diploid offspring is one in which the mother produces a series of eggs that are provisioned, fertilized, and scattered, without subsequent maternal care. Each gene in the mother has an equal probability of being present in each offspring—determined by a flip of the meiotic coin— and the quantity of yolk received cannot be influenced by genes expressed in offspring because provisioning is completed before meiosis. Genes of the mother can do no better than produce the size and number of eggs that maximize the mother's expected lifetime reproductive success. In a simple model in which eggs are produced sequentially on a production line from a fixed quantity of resources, maternal fitness is maximized when each egg is supplied with an amount of resources such that the marginal benefit (δB) that an offspring would gain from a little bit extra committed to its egg would equal the marginal cost of these resources (δC) to another offspring that develops from an egg at the end of the line. Marginal costs and marginal benefits are given equal weight because a gene in the mother has the same chance of being present in either offspring.

The relationship becomes more complex if offspring receive postzygotic maternal care, because the amount of care received can be influenced by genes expressed in offspring. A gene expressed in the current offspring gains the full marginal benefit of extra resources received from the mother but has only a probability r of being present in the offspring that experiences the marginal cost. Therefore, genes expressed in offspring will favor receiving extra resources as long as $\delta B > r\delta C$, whereas genes expressed in the mother will favor terminating investment once $\delta C > \delta B$. Thus, parent–offspring conflict exists whenever $\delta C > \delta B > r\delta C$ (Trivers 1974; Haig 1992a). This conflict arises from the difficulty of making binding agreements. Even though genes in a parent would agree among themselves to terminate investment in each offspring when $\delta B = \delta C$, the agreement is generally unenforceable once genes find themselves in offspring. All genes would do better if offspring demanded less, but unilateral restraint will be exploited.

The probability r is a half for maternal genes expressed in offspring, whereas r will generally be less than half for paternal genes. This is because the offspring that gains a marginal benefit from extra maternal resources may have a different father from the offspring who suffers the marginal cost. Therefore, paternal genes in offspring are predicted to make greater demands on mothers than are maternal genes in the same offspring. Such conditional strategies are made possible by genomic imprinting, which causes genes to have different patterns of expression depending on whether they spent the previous generation in a male or female germline (Moore and Haig 1991). For example, during murine development, *insulin-like growth factor 2* (*Igf2*) is expressed from the paternal allele and the maternal allele is silent, whereas the *insulin-like growth factor 2 receptor* (*Igf2r*) has the opposite pattern of expression. Mice with an inactivated paternal copy of *Igf2* are 60 percent normal size at birth, whereas mice with an inactivated maternal copy of *Igf2r* are 20 percent larger than normal at birth. Birthweight is normal in mice that inherit the inactivated genes from the opposite parent (DeChiara, Robertson, and Efstratiadis 1991; Lau et al. 1994). *Igf2r* has been proposed to function by degrading the products of *Igf2* (Haig and Graham 1991). Thus, these genes employ a conditional strategy of "Make greater demands when paternally derived than when maternally derived."

The relatedness asymmetry between maternal and paternal genes is maximal for half-siblings, but most kinds of relatives will have different degrees of maternal and paternal relatedness. These patterns can be quite complex. Consider a hypothetical social system in which males disperse and females remain in their natal group. If all offspring within a group are fathered by a single male who maintains a monopoly on sexual access to the females until he is displaced by a new unrelated male, female group members of different ages will be closer relatives on the maternal side than on the paternal side because of female philopatry, whereas offspring of the same age will be either full-siblings or paternal half-siblings. It is not known whether genes have evolved conditional strategies that take account of such patterns of relatedness.

All the well-studied examples of genomic imprinting so far appear to be simple conditional strategies of the form "Do one thing when maternally derived and something else when paternally derived." More complex conditional strategies are logically possible—for example, "Do one thing when derived from an egg; something else when derived from a sperm of a resident male; and something else when derived from a sperm of a cuckolding male"—but whether such logical possibilities are actually realized depends on costs, benefits, and the existence of appropriate mechanisms. In the social system of the previous paragraph, a gene's probability of being present in other female group members increases as it passes through successive female germlines. One could imagine a gene subject to a cumulative imprinting effect that was reset to zero every time the gene passed through a male germline.

Reprise

Our intuitive concept of the genetic boundaries of an organism approximates the membership of a coreplicon (that is, of a set of genes that are transmitted by the same rules). A coreplicon evolves as a unit with common goals because its members benefit from the same outcomes, whereas the genes of different coreplicons can have conflicting interests. Thus, viral genes inserted into a bacterial chromosome are distinguished from "true" bacterial genes because of their alternative mode of transmission. They can be mobilized to outreplicate their companions, package themselves into resistant viral particles, and then burst out of their dying host. The relationship between coreplicons need not, however, be strictly adversarial. One coreplicon can obtain a good from another by trade, as well as by theft, but with room for haggling over the price. A coreplicon functions as a commonwealth, without an internal market. Its members thus avoid the transaction costs of finding buyers and of learning the prices and quality of the goods on offer, and are protected from hucksters and frauds in the marketplace (Coase 1993).

Bacterial cells usually contain a small number of coreplicons (sometimes only one). Recombination between bacterial chromosomes is rare and, when it occurs, is a substitutional process in which one gene (a winner) is substituted for another (a loser). By contrast, eukaryotic recombination is much more frequent, and is a segregational process in which genes are swapped between chromosomes, without winners and losers. The coreplicon is no longer a group of nonrecombining genes who cooperate because their long-term fates are intimately bound; rather, it becomes a group of temporary associates who obey the same rules and who gain an equitable division of resources when their ephemeral partnership is dissolved. High-frequency recombination creates a market (of a sort) for currently favored team players.

The reasons for the eukaryotic sexual cycle of gametic fusion, recombination, and meiotic segregation remain somewhat unclear, but the process probably enhances individual genes' chances of long-term survival in a changing selective environment. Recombination creates *relatives*—genetic collectives that share some, but not all, of their genes. Interactions with relatives are a potential source of internal dissension within the collective, because some members of the collective can gain at the expense of others, for example, by sabotaging other members' gametes or by favoring some offspring over others. High levels of recombination can be a partial solution to the conflicts created by lower levels of recombination, because randomizing devices disrupt the "cabals of the few."

3 The "Gene" Meme

The final chapter of *The Selfish Gene* (Dawkins 1976) explored the analogy between genetic and cultural evolution. Cultural traits, Richard Dawkins suggested, evolve by a process of natural selection in which there is preferential proliferation of traits with properties that promote their own transmission. "We need a name for the new replicator," he wrote, "a noun which conveys the idea of a unit of cultural transmission, or a unit of imitation. *'Mimeme'* comes from a suitable Greek root, but I want a monosyllable that sounds a bit like 'gene.' I hope my classicist friends will forgive me if I abbreviate mimeme to *meme*." Dawkins concluded his discussion with: "However speculative my development of the theory of memes may be, there is one serious point which I would like to emphasize once again. This is that when we look at the evolution of cultural traits and at their survival value, we must be clear *whose* survival we are talking about." He entertained the possibility "that a cultural trait may have evolved in the way that it has, simply because it is *advantageous to itself*."

The "meme" has exhibited admirable powers of replication and persistence in the thirty years since its conception, but its cultural spread pales before that of the monosyllable it was

chosen to imitate. In the first half of this chapter, I will consider
the diverse meanings that have become associated with that simple meme, the "gene." Not every scientist means the same thing
when they refer to a gene, and these differences in nuance can
be a source of confusion. In particular, I will discuss Dawkins's
explicit definitions of the selfish gene and, in the guise of *the
strategic gene*, propose what I believe to have been Dawkins's
implicit definition. We can think of the changing and diversify-
ing concepts of the gene as an example of memetic evolution.
The second half of this chapter will use the discussion of the
"gene" in the first half to illuminate the status of the "meme" as
a putative replicator subject to cultural selection.

The Danish plant breeder Wilhelm Johannsen (1909, 124)
introduced *das Gen* (plural *die Gene*) into the German language
as a shortening of Darwin's "pangene" to replace the ambiguous,
polysemic *Anlage* (unit). He then used the singular "gene" in an
English-language address to the American Society of Naturalists
in December 1910, published the following year. Johannsen's
intent in this address was to take issue with the common "conception that the personal qualities of any individual organism
are the true heritable elements or traits!" The rediscovery of Mendelism had shown that "the *personal qualities* of any individual
organism do not at all cause the qualities of its offspring; but the
qualities of both ancestor and descendant are in quite the same
manner determined by the nature of the 'sexual substances'—
i.e., the gametes—from which they have developed. Personal
qualities are then *the reactions of the gametes* joining to form a
zygote; but the nature of the gametes is not determined by the
personal qualities of the parents or ancestors in question." Thus,
Johannsen (1911) made a crucial distinction between *phenotype*
(observable traits) and *genotype* (heritable factors).

In Johannsen's view, the mistaken notion of the inheritance of personal qualities was reinforced by the persistence of an outdated vocabulary. "It is a well-established fact that language is not only our servant, when we wish to express—or even conceal—our thoughts, but that it may also be our master, overpowering us by means of the notions attached to the current words. This fact is the reason why it is desirable to create a new terminology in all cases where new or revised conceptions are being developed. . . . Therefore I have proposed the terms 'gene' and 'genotype' and some further terms, as 'phenotype' and 'biotype,' to be used in the science of genetics. The 'gene' is nothing but a very applicable little word, easily combined with others, and hence it may be useful as an expression for the 'unit-factors,' 'elements' or 'allelomorphs' in the gametes, demonstrated by modern Mendelian researches" (1911, 132).

From this beginning, Johannsen's "gene" has had an illustrious history, as have "genotype" and "phenotype" but not "biotype." But "gene" itself conveys little information, consisting as it does of only four letters and a single syllable when spoken. The factors accounting for its success as a meme are probably those identified by Johannsen—that it was "a very applicable little word, easily combined with others"—and the historical contingency that the word was used to represent a set of ideas and concepts that had high memetic fitness. If "gene" is a meme, it is a rather uninteresting one. The interesting memes are the shifting concepts of the units of inheritance for which gene was a convenient label. The memetic history of "gene" is interesting only insofar as it provides a marker for the propagation of these more amorphous ideas and concepts as they have undergone constant reformulation.

"Gene" has never had a single meaning, but has always had different meanings for different people, and often different meanings for a single person, depending on context. For each person who added "gene" to their vocabulary, the word had a meaning that was derived from explicit definitions either read or heard, inferences from how the word was used, and reformulations of the concept within their own minds. This private definition of "gene" was then translated into new definitions and new uses in conversation and writing that were perceived by other minds and incorporated into new private definitions. My intention in stating the obvious is to point out what must be true of most memetic transmission: there is some degree of continuity in the propagation of ideas from mind to mind, but it lacks the high fidelity of the propagation of genes from generation to generation.

Johannsen, of course, had his own conception that he wished to convey to others. "As to the nature of the 'genes' it is as yet of no value to propose any hypothesis; but that the notion 'gene' covers a reality is evident from Mendelism. . . . We do not know a 'genotype,' but we are able to demonstrate 'genotypical' differences or accordances. . . . Genotypes can be examined only by the qualities and reactions of the organisms in question" (1911, 133). Genes were known by their phenotypic effects. Johannsen was dismissive of attempts to localize genes. "The question of *chromosomes* as the presumed 'bearers of hereditary qualities' seems to be an idle one. I am not able to see any reason for localizing 'the factors of heredity' (*i.e.*, the genotypical constitution) in the nuclei. The organism is in its totality penetrated and stamped by its genotype-constitution. All living parts of the individual are potentially equivalent as to genotype-constitution" (154).

Johannsen's "applicable little word" soon gained wide currency among geneticists, especially among those who believed, contrary to Johannsen, that the gene corresponded to a physical structure on chromosomes. One might say there had been memetic recombination that attached "gene," as a label, to an alternative concept of the unit of inheritance. Supporters of the chromosomal theory however continued to define the gene operationally as that which was responsible for a heritable phenotypic difference. Alfred Sturtevant, one of the first to map genes to chromosomes, commented: "We can . . . in no sense identify a given gene with the red color of the eye, even though there is a single gene differentiating it from the colourless eye. So it is for all characters. . . . All that we mean when we speak of a gene for pink eyes is, a gene which differentiates a pink eyed fly from a normal one—not a gene which produces pink eyes *per se*, for the character pink eyes is dependent on the action of many other genes" (1915, 265).

Much of twentieth-century experimental genetics was engaged in making inferences about the physical nature of genes from observations of differences in the physical characteristics of organisms. These studies led to a definition of the gene as a stretch of DNA that was responsible for specifying the amino acid sequence of a protein. Thus, the operational definition of a gene—a gene is known by its effects on outward form—began to shift as genes came to be viewed as tangible elements with defined chemical properties. The existence of a gene is now often inferred from properties of a DNA sequence without any information about the gene's phenotypic effects and without any observation of differences among DNA sequences. But this definition of the gene, as a protein-encoding stretch of DNA, is more recent than its definition as that which is responsible for a phenotypic difference, and it is not surprising that the modern molecular definition has not fully supplanted the older operational definition.

Experimental geneticists invoke genes to explain *observed* phenotypic differences. A pink-eyed fly differs from a red-eyed fly because the former possesses a gene for pink eyes inherited from both parents whereas the latter has inherited at least one gene for red eyes. In a similar way, evolutionary biologists often invoke genes to explain *hypothetical* phenotypic differences in an attempt to understand the nature of adaptation by natural selection. An ornithologist might wish to understand why males of some species help to raise offspring (dads) while males of other species put all their efforts into seeking additional copulations (cads). She might posit a gene for being a cad and ask under what circumstances it would invade a population in which most males behave as dads. To paraphrase Sturtevant, all that we mean when we speak of a gene *for* being a cad is a gene that differentiates a cad from a dad—not a gene that produces caddish behavior *per se*, for caddishness results from the action of many genes.

The use of genes to explain both differences among individuals and invariant features of organisms has led to an unfortunate confusion in public discourse over claims that genes cause behavior. Consider a contentious example. Behavioral geneticists are interested in differences among individuals and look for genetic factors that might explain why some people, but not others, engage in violent acts. Perhaps violent offenders are unable to control their impulses because they carry a mutation in the gene encoding the enzyme monoamine oxidase. The explicit comparison is between the behaviors of individuals with different variants of the gene. Evolutionary psychologists, on the other hand, are interested in species-typical behaviors that they view as adaptations to enhance the survival and reproduction of individuals. Therefore, they seek explanations for why we have evolved a genotype that makes us more likely to engage in violence under some circumstances, but not others. Perhaps young men are predisposed to violent behavior when they control few resources in societies in which there are large differences between rich and poor. The implicit comparison is between the reproductive fitness of individuals in the present world (or in a past environment in which the behaviors were adaptive) and the reproductive fitness of individuals in alternative worlds in which genes respond differently to the environment. Thus, behavioral geneticists ascribe the observed difference between offenders and nonoffenders to a genetic difference, whereas evolutionary psychologists ascribe the same difference to environmental factors. Yet both are castigated as "genetic determinists" by those who reject biological explanations of human behavior.

How then did Dawkins define the eponymous protagonist of *The Selfish Gene*? Dawkins (1976) recognized that there is "no universally agreed definition of a gene. Even if there were, there is nothing sacred about definitions. We can define a word how we like for our own purposes provided we do so clearly and unambiguously. The definition I want to use comes from George Williams. A gene is defined as any portion of chromosomal material that potentially lasts for enough generations to serve as a unit of natural selection" (1966, 30). Thus, the gene could be a longer or shorter unit than the protein-encoding gene recognized by molecular biologists. When defined in this way, Dawkins believed that the gene must be recognized as "the fundamental unit of natural selection, and therefore the fundamental unit of self-interest" (35).

Dawkins's recognition of the gene as the unit of selection has not met with universal acceptance, partly because different scientists have different implicit definitions of the gene. David Sloan Wilson and Elliot Sober (1994), for example, have argued that the gene deserves no special status because it is merely the lowest level of a nested hierarchy of units of selection: genes, cells, individuals, groups, species. In this view, genes are nested within cells, cells within individuals, and individuals within groups; and natural selection can act at all levels of the hierarchy. Thus, there may be adaptations for the good of individuals and groups as well as of genes. The placement of the gene at a level of the hierarchy below the cell implicitly defines the gene as a material object located within cells. But this is not Dawkins's concept: "What is the selfish gene? It is not just one single physical bit of DNA . . . it is *all replicas* of a particular bit of DNA distributed throughout the world. . . . The key point . . . is that a gene might be able to assist replicas of itself which are sitting in other bodies. If so, this would appear as individual altruism but it would be brought about by gene selfishness" (1976, 95). For Wilson and Sober, the gene is a *material* object that resides within cells, whereas for Dawkins the gene is a piece of *information* distributed across multiple levels of Wilson and Sober's hierarchy. I cannot resist suggesting the usit—pronounced "use it"—as an applicable little term to represent the disputed unit-of-selection whatever that may be or mean.

The continued debate between gene-selectionists and hierarchical-selectionists identifies an ambiguity in meaning even when a gene is defined as a rarely recombining stretch of DNA. A gene could refer to the group of atoms that is organized into a particular DNA sequence—each time the double helix replicates, the gene is replaced by two new genes—or it could refer to the abstract sequence that remains the same gene no matter how many times the sequence is replicated. In the previous chapter, I called these concepts the *material gene* and the *informational gene*. Dawkins referred to something like the informational gene when he described the selfish gene as "*all replicas* of a particular bit of DNA," but I believe that he neither wanted nor intended this definition. If all humans came to share the same DNA sequence, the theory of the selfish gene would not predict universal benevolence. A selfish gene does not "care" about all replicas of its sequence, but only about some of its replicas in a smaller group of related individuals. The reason why is tied up with the dynamics of genetic replicators.

Mutations create new informational genes that are modifications of existing informational genes. Each mutation occurs as a change to a single material gene. Therefore, the mutation must, at first, be a rare variant in the gene pool and its material copies will interact with each other only in cells of the same body or bodies of close relatives. Such interacting groups of material genes correspond to the *strategic genes* (or units of adaptive innovation) of the previous chapter. For such a mutation to increase in frequency by natural selection, its phenotype must benefit the transmission of its copies from this small group of individuals relative to the transmission of alternative informational genes by other small groups of related individuals. If the new informational gene increases in frequency because of this phenotype, strategic genes of its type will encounter each other in interactions between distantly related individuals but will still promote the nepotistic phenotypes that ensured their informational gene's success when rare. For this reason, a mother will favor her own child over a twelfth cousin even though the child and twelfth cousin are identical for most of their informational genes. Genetic competition is predicted in populations of genetically identical individuals.

A material gene has dual roles. It can be *expressed*—that is, its sequence can be transcribed into a messenger RNA that is translated into a protein—and it can be *replicated* as copies of itself. The essence of adaptation by natural selection is that a material gene's phenotypic effects, how it is expressed, influence the probability that the material gene, or its replicas, will be copied. The extent of the strategic gene is determined by how many replication cycles separate the material genes responsible for a phenotypic effect from the material genes that benefit from an increased probability of being copied. Thus, the strategic gene is not a fixed entity but one that can evolve to encompass more, or fewer, material copies of an informational gene.

Take for instance a large, well-mixed population of single-celled phytoplankton. Once a cell divides, the daughter cells separate and never interact again, except by chance. Each material gene is subject to selection solely on how its own expression influences its own replication. In this case, the strategic gene is limited to a single material gene. Now consider a cod fish in which multiple individuals of both sexes spawn simultaneously. A sperm and egg fuse to form a zygote that develops into a large multicellular individual that may itself contribute eggs or sperm to zygotes of the next generation. Zygotes become widely dispersed by ocean currents so that there are no preferential associations among kin. A single material gene in a zygote gives rise to replicas in all the cells of an adult cod. Material genes in a cod's heart and brain never replicate, yet their expression promotes the replication of their replicas in the cod's gonads. In this example, the strategic gene is spread throughout the body of a single fish. Finally, consider a beehive. Material genes that are expressed in the hive's sterile workers promote the replication of their copies in the ovary or sperm storage organs of the queen bee. So in our third example, the material copies of the strategic gene are spread among the members of the hive. When Dawkins discusses selfish genes, it is in the sense of the strategic gene that his genes should be understood.

Dawkins proposed that memes play a role in cultural evolution that is analogous to the role played by genes in biological evolution. If so, memes should display features that promote their own replication. Such features could be interpreted as adaptations "for the good of" the meme itself. The remainder of this chapter will interrogate the analogy between genes and memes. I will employ a vague definition of a meme as "a mental item that is borrowed from one person and passed on to another." There are many things that could be considered to be memes, but my focus will be on the transmission of ideas, and I will use the meme of the "gene" in counterpoint.

Rather than asking directly who benefits from memetic transmission, let's consider who benefits from communication. Many acts of communication are committed because a sender wants to produce some change in a receiver. Such acts can be considered propaganda, from the Latin for propagation. An item of propaganda, a *propagandum*, is a device designed by a propagandist to achieve a change in the actions of a receiver. The propagandum has served the *purpose of the propagandist* if the receiver acts in the desired manner. To achieve this purpose it isn't necessary that a receiver pass the propagandum on to others. If there is no chain of transmission, then the propagandum does not qualify as a meme. Its effects are not advantageous to the propagandum, but only to the propagandist.

Sometimes a propagandum is designed to be passed from one receiver to another because this increases the propagandum's efficiency as an agent of mass persuasion. If the propagandist is successful in achieving ongoing transmission, the propagandum then qualifies as a meme. The propagandum serves its designer's ends if receivers act in some desired manner and if receivers pass the propagandum on to others to affect their behavior. The features that promote the propagandum's transmission benefit the propagandist but can also be said to benefit the propagandum considered as a meme. But the features that effect a change in the behavior of receivers, while they benefit the propagandist, need not benefit the meme.

A propagandist's designs may misfire. A propagandum could fail to achieve the propagandist's ultimate purpose of changing behavior, but could succeed in the subsidiary purpose of being propagated from mind to mind; or, the propagandum could continue to propagate from mind to mind after it no longer serves the propagandist's goals. Once a chain of memetic transmission exists, there is no selection on a propagandum to serve its original designer's ends, although the propagandum, as meme, may continue to serve these ends if the fidelity of transmission is sufficiently high.

Each step in a chain of transmission is an act of *selection*, insofar as the transmitter chooses to transmit one meme rather than another (or no meme at all). We can think of any feature of a meme that has predisposed successive transmitters to "want" to pass it on as an adaptation of the meme to enhance its own transmission: such adaptations can appeal to either the conscious motivations or the unconscious motivations and biases of transmitters; and such adaptations can be intended features, consciously selected by a propagandist, or they might be unintended features that arise from the interaction of "random" mutation with differential replication during the chain of transmission. Thus, the adaptive features of memes may be the products of "intelligent design," "natural selection," or a combination of the two.

Whose interests then are served when a meme is transmitted? We can look at this from the perspectives of individuals or memes. From the first perspective, we need to consider the interests of the transmitters at each step in the chain. If an individual consciously chooses to transmit a meme, then the meme must serve some perceived interest of theirs. I write *perceived interest* because individuals can be mistaken about what will promote their true interests. For example, a meme may be a propagandum serving the *actual* interests of some other individual earlier in the chain. (By the interests of an individual, I mean here their own self-defined goals in life.)

Does taking the second perspective, of viewing culture through the lens of a meme's metaphorical interests in its own transmission, contribute anything that could not be obtained from the first perspective? A meme's-eye view could be justified if it were shown that there are features of memes that promote a meme's own interests without serving the interests of any of the meme's transmitters. Such features might appeal to quirks of nervous systems that are better considered as unconscious biases rather than sources of personal motivation. A preference for a meme's-eye view might also be defended if the features that make a meme likely to be passed on had accumulated in many steps over the course of memetic transmission.

Johanssen introduced "gene" to clarify the distinction between genotype (gene) and phenotype (trait). Can a similar distinction be made for a science of memetics? There are two principal kinds of things we observe that provide evidence about the nature of memetic transmission. The first are communication acts including sounds, texts, actions, and artifacts. The second are insights from introspection when we register a communication act, when we integrate the content of a communication act into our private set of concepts, and when we emit communication acts. Introspection may be an unreliable guide because unconscious aspects of our motivations are hidden and our conscious perceptions may be partial, inaccurate, and misleading. Communicative acts appear closer to the concept of genotype (things transmitted) whereas the conscious and unconscious effects of these acts on our internal state appear closer to phenotype (effects that influence what is transmitted). In the history of genetics, the phenotype was observed and the genotype inferred. But this relation is reversed for memetics. Memes are observed, but their effects are inferred.

The phenotype–genotype distinction works fairly well for genes, but there are many unresolved problems in its application to memes. For example, let's suppose that there is a body of lore preserved and updated by the Medieval Guild of Propagandists about what techniques are effective in changing public opinion, and that this lore is passed from master to apprentice. An apprentice uses techniques of proven efficacy to design propaganda, and then a propagandum's success in public persuasion influences whether the apprentice passes on the technique to his own apprentices when he becomes a master. From the perspective of the techniques as memes, propaganda are meme-products that influence a technique's probability of transmission, but these items may also function as memes in their own right. A propagandum may be both "memotype" and "phemotype."

The gene has a material definition in terms of a DNA sequence that maintains an uninterrupted physical integrity in its transmission from generation to generation. Memes also have a physical form in their transmission from one individual to another, sometimes as sound vibrations, as text on paper, or as electronic signals relayed through a modem. When these "outward" forms of a meme are perceived, they elicit changes in a nervous system that constitute the meme's "cryptic" form. The material basis of the cryptic form is probably unique to each nervous system colonized by the meme. Memetic replication, then, has nothing like the elegant simplicity of the double helix.

If the material form of memes is problematic, might it be more appropriate to define memes purely in terms of information? But what are the memes in our evolving concepts of "the gene"? These concepts have been reformulated and recombined with other ideas at each step in the chain of transmission. How can one identify the "nuggets" of ideas that remain unchanged during this process and thus persist "for enough generations to serve as a unit of natural selection"? Dawkins argued that "selfishness is to be expected in any entity which deserves the title of a basic unit of natural selection." His definition of the gene qualified as such an entity because it possessed three properties that "a successful unit of natural selection must have . . . longevity, fecundity, and copying fidelity." Unlike his careful definition of the gene, Dawkins was somewhat vague about the definition of a meme, simply stating that this was a "unit of cultural transmission, or unit of imitation." Is there some way to define a meme so that it possesses the properties that would qualify it as a unit of natural selection and hence deserve the "selfish" label?

Consider the 215 pages of the first edition of *The Selfish Gene*. Dawkins's slim volume contains many ideas influenced by older texts and itself has influenced ideas expressed in newer texts (including this one). Can *The Selfish Gene* be parsed into a set of selfish memes each displaying longevity, fecundity, and copying fidelity? But do we similarly parse the genome into discrete genes? Dawkins's definition of the gene did not specify boundaries between genes. The gene was a piece of chromosome sufficiently short to last long enough, without recombination, to function as a unit of selection. But this definition meant that there were many different, overlapping ways in which a chromosome could be divided into genes. Could the same approach work for memes?

Dawkins's principal interest was in the phenotypes of organisms rather than of genes. Just as he failed to specify precisely how to divide a chromosome into genes, he did not specify how to divide the phenotype up into individual adaptations due to individual genes. I believe this approach was justified for his purposes. As long as all parts of the genome have the same rules of inheritance, what is good for one part of the genome is good for all parts of the genome, and the genome itself can be considered as an adaptive unit. Highly complex adaptations require a long genetic text. There are two widespread solutions to this problem that one can call the asexual and the sexual solution. In the asexual solution, an entire genome replicates as a unit and does not recombine with other genomes. Thus, the whole genome behaves as a single Dawkinsian gene. In the sexual solution, two entire genomes come together for some length of time, then separate into two new genomes after exchanging interchangeable parts, with each new genome receiving one of each part. The sexual genome is an ephemeral collective of many Dawkinsian genes, but the rules of Mendelian inheritance ensure that what is good for one part is good for all, at least for the time that the genes are temporarily associated. (I leave to one side the complexities that arise when the "rules" are broken and there is conflict within the genome.)

Neither the sexual nor the asexual solution seems to apply to most complex memetic "texts." Ideas recombine freely to generate each new text, and there is no well-defined exchange of interchangeable parts. One idea can be adopted from a text and the remainder abandoned. Therefore, the adaptations of memes will be adaptations for the good of the individual rarely recombining ideas. Some of these ideas may be so simple—for example, the idea that the gene is a part of a chromosome—that they can exhibit few, if any, adaptations for their own transmission. I see little value in treating such ideas as selfish, just as there is little value in treating a single nucleotide as selfish. Such ideas serve the utility of propagandists or (perhaps) of larger nonrecombining meme complexes into which they become incorporated. The place to look for sophisticated adaptation and selfishness will be in coherent ideologies, large "asexual" meme complexes that are transmitted as a unit with high fidelity of transmission. Richard Dawkins would identify the world's great religions as the prime examples and would argue that the free recombination of ideas is important if ideas are to serve our ends rather than their own.

These are some of the problems I see in defining memes and thinking of memes as selfish. And yet, in the four decades since I first read *The Selfish Gene*, no section of the book has stayed more in my mind than its final chapter. In the intervening period, I have disseminated the selfish meme many times in conversation, and here I am spreading the meme in print. The meme of the "meme" is a tenacious beast, at least for those minds that are vulnerable to its charms. The current chapter is a work of propaganda. I wish to communicate ideas that I hope will influence your own concepts of genes and memes. If I am effective, you may pass on these ideas in modified form to others. In pursuit of these ends, I have crafted phrases to *grab your attention*, and have worked on clarifying concepts in my own mind. This process has involved testing numerous alternatives against the standards of what I think will be effective and what will form a coherent whole with the rest of the chapter. I have read and revised my text innumerable times. I think of the final product as expressing my own intentions rather than serving the ends of the ideas it tries to communicate. But am I fully autonomous in this process? Many ideas have competed for inclusion during the course of writing, but only some have made it into a final version that has nothing like the form and content that I intended when I first sat down to write. It is only in retrospect that I know what I mean. The final version contains the ideas that have *grabbed my attention*. It has sometimes seemed that they have been using me for their ends. What fraction of these ideas are my own and what fraction have been borrowed from others? The web of intellectual influence is complex, and it is unclear whether I ever have a truly original idea.

Darwin's disciple, the Oxford biologist-*cum*-psychologist George Romanes, wrote in *Darwin, and After Darwin*:

Quite apart from any question as to the hereditary transmission of acquired characters, we have in this *intellectual* transmission of acquired experience a means of *accumulative* cultivation quite beyond our powers to estimate. For . . . in this case the effects of special cultivation do not end with the individual life, but are carried on and on through successive generations *ad infinitum*. . . . [In] this unique department of purely intellectual transmission, a kind of non-physical natural selection is perpetually engaged in producing the best results. For here a struggle for existence is constantly taking place among "ideas," "methods," and so forth, in what may be termed a psychological environment. The less fit are superseded by the more fit, and this not only in the mind of the individual, but through language and literature, still more in the mind of the race. (1895, vol. 2, 32)

Richard Dawkins exhorted us to ask: "fit" in what sense and for whom?

4 Differences That Make a Differance

My contemplation of the many meanings of "gene" identified a subtle ambiguity in usage. The classical gene was a hypothetical entity that was invoked to explain an observed difference. Genes were identified by the heritable differences they caused. Thus, the *gene* was implicitly defined as a difference because only a difference can cause a difference. On the other hand, genes were also thought of as physical *things*. The rival concepts of *genes as differences* and *genes as things* each have their uses but, when we address questions of adaptation and natural selection, it is the *gene as heritable difference-maker* that is the relevant concept. Natural selection "chooses" among alternatives by the criterion of the differences they cause. These alternatives are differences that make a differance. There can be no differential replication without a difference, just as there can be no choice without an alternative, and there can be no cumulative change without heritability of the causes of difference.

Gene-selectionism is the conceptual framework that views genes as the ultimate beneficiaries of adaptations and organisms or groups as means for genes' ends. Rival conceptual frameworks exist. Multilevel selection theory views genes as the lowest level of a nested hierarchy in which each level is subject to selection and each level can be a beneficiary of adaptations (Sober and Wilson 1994; Wilson and Sober 1994). Developmental systems theory similarly denies a privileged role for genes in development and evolution. In this framework, many things other than genes are inherited and many things other than genes have a causal role in development. It is the entire developmental system, including developmental resources of the environment, that reconstructs itself from generation to generation (Gray 1992; Oyama 2000; Sterelny and Griffiths 1999). Arguments over the relative merits of the different frameworks can be heated. Some of this argument is substantive but much is semantic. Different frameworks define fundamental terms in different ways. Without close attention to these semantic differences, substantive issues can become obscured by mutual incomprehension.

This chapter explicates the nonstandard definitions of gene, phenotype, and environment used by one particular gene-selectionist (myself) in some particular contexts. Although I strive for precision in definition, no definition will be unassailable. Natural selection constantly undermines rigid definitions because it is a process by which things of one kind become things of a different kind. Definitions themselves evolve and words acquire different meanings in different contexts. No language is unobjectionable if a hostile reader can choose how to interpret terms. Mutual understanding should be facilitated by a clarification of how terms are used, but uniformity of definition is neither achievable nor desirable.

My intent is to clarify how central concepts should be defined to achieve a consistent gene-selectionism, not to argue that gene-selectionism is superior to other frameworks. Although I am a gene-selectionist by predilection, I have respect for and sometimes use multilevel selection theory. It is consistent and coherent. I also respect developmental systems theory, especially in its account of development. I see these frameworks as heuristic devices for thinking about evolutionary questions. These alternative frameworks may be better suited to particular temperaments and particular questions.

Phenotypes

Phenotypes have traditionally been defined as properties of organisms and thus must be redefined if phenotypes are to be considered properties of genes. *A gene's effects are its phenotype* (Dawkins 1982, 4). In this definition, an effect is simply a difference from what would be observed in the absence of the gene or in the presence of a variant gene, other things being equal. A gene considered in isolation does not have a phenotype. All assignments of phenotypes are based on comparisons (Bouchard and Rosenberg 2004). These comparisons are implicit in the measurement of genic fitness as a change in relative allele frequency. Thus, a gene's phenotype depends on the implicit or explicit alternative with which it is compared. For some evolutionary questions, comparisons are made between existing variants of a known sequence. For other questions, an existing gene might be compared to a hypothetical alternative, or a hypothetical gene and hypothetical alternative might be posited as making a specified difference in the world without tying the difference to a particular genetic locus.

Consider a sequence variant that has become fixed in a population by a selective sweep. The variant allele was initially present in a single copy in a single cell. Early in the sweep, the allele's increase in frequency was determined by its effects relative to the previously established allele. Once the sweep is complete, the allele maintains its high frequency, or is itself replaced, depending on its effects relative to new mutations. Some mutations will result in complete loss of function. In this case, the phenotype that is subject to selection is the difference between a functional and nonfunctional allele. Other mutations will cause expression in a new cell type, or a change in alternative splicing, or a change in promoter activity, and so on. For each kind of mutation, there is a different difference under selection.

The definition of phenotype as a gene's effects changes how environmental and genetic factors are conceptualized. Phenotype is no longer seen as a sum of genetic and environmental influences (plus interaction terms). All phenotypes are effects of genes, but a gene's phenotype may encompass different effects in different environments (the gene's repertoire of effects) and includes the gene's effects on the environment. If there are differences among organisms that are uninfluenced by genes, then these differences are part of no gene's phenotype. If a frog has a leg amputated by some purely random event, then the absence of the limb is not a phenotype, although how the frog copes, or fails to cope, with the amputation may be part of the phenotype of many genes. We are all interactionists. A definition of phenotype in which all effects are ascribed to genes, but genes' effects may vary among environments, has much to recommend it if it helps us move beyond sterile debates about nature versus nurture.

A gene achieves its effects by interacting with proteins, RNAs, DNA, and other molecules. The other molecules with which it interacts can be conceptualized as part of the gene's environment. Of particular importance are a gene's interactions with RNA polymerases in the process of transcription. But a gene's phenotype need not be mediated solely by the coding and noncoding RNAs transcribed from its sequence. DNA sequences themselves can adopt multiple conformations depending on conditions in the nuclear environment and these conformations can influence whether the gene is transcribed. For example, the 5′-flanking region of human γ-globin genes responds to low pH by forming an intramolecular triplex in which a purine-rich strand inserts into the major groove of the contiguous double-helix, leaving an unpaired pyrimidine-rich strand. Point mutations that destabilize this structure are associated with hereditary persistence of fetal hemoglobin (Bacolla et al. 1995). Thus, the pH-dependent ability to fold back upon itself and thereby inhibit its own transcription can be considered part of the phenotype of the γ-globin gene. As another example, the imprinted H19 gene interacts directly with imprinted regions on other chromosomes and influences when in the cell cycle these other genes are replicated (Sandhu et al. 2009).

Genes have been viewed as catalysts that facilitate chemical reactions but are not changed by those reactions. Much recent attention, however, has focused on chemical reactions that alter genes. For example, a DNA sequence may be "epigenetically" modified by its interaction with methyltransferases. If alternative DNA sequences exist that are not subject to methylation, then methylation is part of a gene's phenotype compared to those alternatives. If a methyl group, once attached, is faithfully inherited and more or less permanent, then a methylated and unmethylated sequence can be considered alternative genes compared to each other. But, if a genetic lineage switches back and forth between methylated and unmethylated states, then the ability to switch between states (the gene's norm of reaction) can be interpreted as part of the gene's phenotype compared to alternative alleles that do not undergo switching (Haig 2007).

Functions and Side Effects

A gene may have effects that influence the probability that it will be replicated. Genes that promote their own replication will be perpetuated, whereas alternative genes that are less effective replicators will be eliminated. The *effects* of a DNA sequence may thus be included among the *causal* factors that account for the presence of the sequence in a gene pool. It is this causal feedback between genotype and phenotype—when combined with a source of genetic novelty (mutation)—that explains how a purposeless process (natural selection) can produce purposeful structures and functions (adaptation). The environment selects among phenotypes and *thereby selects among genes*. By this means, gene sequences come to embody and represent "information" about what has worked in the environment (Frank 2009; Shea 2007).

The effects of a gene can be classified as either *functions* (effects that are beneficial for the gene) or *side effects* (effects that are neutral or harmful for the gene). Effects are judged as functions or side effects by their average contribution to replication over many occurrences not from a single occurrence. A gene's functions consist of those of its effects that have contributed, however indirectly, to its own transmission from past generations. Insofar as the future repeats the past, such functions will contribute to the gene's transmission to future generations. All effects of a gene comprise its phenotype and are subject to selection, but only those effects that promote a gene's replication comprise its functions. There is selection *for* a gene's functions but selection *of* its side effects. If selection is to choose a harmful effect then it must be associated with an even greater beneficial effect (a selective trade-off).

Functions are the adaptations of genes. For an effect to qualify as a function, variant genes must have been eliminated in the past because they lacked the effect, and, if the effect is to remain a function, such variants must continue to be eliminated when they arise. A gene's effects can change when the environment changes, and a given effect can shift its status from function to side effect or the reverse. There is no inconsistency in saying that an effect was once a function but is so no longer. Teleological language is appropriate when referring to the functions of genes because functions are final causes. They are both causes of a gene's persistence and effects of the gene.

Environments

A gene's environment encompasses all factors that are shared with the alternative against which the gene's effects are measured. It contains not only factors external to the cells and bodies of organisms, but also (and more immediately) these cells and bodies themselves. A body can be viewed as the collectively constructed niche of the genes of which it was the extended phenotype. Among the most important parts of a gene's environment are the other molecules with which it interacts. Other genes, even other alleles at the same locus, are parts of a gene's social environment (Fisher 1941; Okasha 2008; Sterelny and Kitcher 1988). Any factor that is experienced by a gene, but not by its alternative, belongs to the gene's phenotype, not its environment.

Genes may have effects that vary in different environments, in adaptive or nonadaptive ways. If alternative genes experience a similar range of environments but exhibit different responses to these environments, then their norms of reaction are phenotypes subject to selection. Genes may also have effects that modify the environment or phenotype of other genes.

Genetic inheritance and phenotypic development can be conceptualized as orthogonal axes (Bergstrom and Rosvall 2011). The vertical axis represents transmission of old information from progenitors to progeny whereas the horizontal axis represents development of each new generation. Relations between genes and environment differ on the two axes. The environment has primacy on the vertical axis. The information of genes comes from the environment via selection among genetic differences. But, on the horizontal axis of public performance, genes and environment interact to create form. Neither has explanatory primacy. On the vertical axis, genes refer to *past* environments, whereas on the horizontal axis, genes interact with the *current* environment.

Genealogical (vertical) and ontogenetic (horizontal) axes are not causally isolated. If a gene makes a selective difference in the current environment, this will be reflected in a changed composition of the gene pool. In any particular generation, entire genomes are selected and one cannot ascribe a selective difference to a particular gene. However, genomes are disassembled and reassembled in each generation by processes of recombination. Therefore, over a series of generations, shorter sequences of DNA are tested against multiple genetic backgrounds. On the ontogenetic axis, the effects of genes are highly nonadditive, because of complex interactions with the environment (which includes other genes), but on the genealogical axis, sustained changes in gene frequency are explained by the average additive effects of smaller parts of the genome (Ewens 2011; Fisher 1941).

Are Genes Dispensable?

To recapitulate, a gene's world can be divided into phenotype and environment. Phenotype is the difference between the alternatives under selection and environment is the sameness. Phenotype consists of those parts of a gene's world that differ from the world of the alternative gene to which it is compared whereas environment contains those parts that are shared with the alternative's world. In this formulation, phenotype and environment are properties of comparisons between things not properties of the things themselves. The environment chooses between the alternative phenotypes and thereby chooses between the genes. Could a simpler story be told in which the environment selects among heritable phenotypic differences without the need to invoke heritable determinants of difference?

The principal reason for invoking determinants is that causality matters. Elsewhere I have considered a model in which heritability of birth weight was high and different birth weights were associated with different probabilities of survival, but there were no fitness differences associated with genetic causes of variation (Haig 2003). Rather, fitness differences were associated solely with environmental contributions to the variance. Perhaps all maternal genotypes produce heavier and healthier babies in better environments. If so, there would be the appearance of directional selection for heavier babies but no response to selection. Perhaps greater environmental perturbations from genotype-specific optimal birth weights result in reduced survival of babies. If so, there would be the appearance of stabilizing selection but no reduction in genetic variance. In both scenarios, birth weight and fitness are correlated, and birth weight is

heritable but fitness is not. Heritability of a trait that is correlated
with fitness is not enough. Natural selection requires heritability
of the *causes* of variation in fitness. The second reason for invoking genes as heritable determinants of difference is that genes
can be considered agents that benefit from the phenotypes they
cause.

Are Genes Countable?

George Williams and Richard Dawkins defined a gene as a rarely
recombining stretch of DNA that is transmitted intact over
multiple generations (Williams 1966, 24; Dawkins 1976, 30).
This definition includes stretches of DNA that are the "context-
sensitive difference makers" of Sterelny and Griffiths's (1999, 87)
definition of a gene but need not be restricted to that category.
The existence of DNA sequences without effects has been presented as a problem for the evolutionary gene concept but the
difficulty evaporates if it is conceded that an evolutionary gene
need not be subject to selection. Nor is there a problem if different DNA sequences have identical effects. One can identify
two genes with the same effects as different genes on the basis
of their sequence. A gene without effects is a gene without a
phenotype. Such genes make no selective difference, although
their relative frequency may change by genetic drift or draft.
Similarly, a gene that has erratic effects on its own replication,
such that there is no average effect, is not subject to selection,
but remains a gene. These are differences that do not make a
difference.

The linear extent of an evolutionary gene can be considered
to be the distance along a chromosome over which genetic differences are correlated. Consider two neighboring DNA segments

X and Y. If P(X) and P(Y) are the relative frequencies of X and Y, and P(XY) is the frequency with which X occurs together with Y, then the distributions of X and Y are statistically nonindependent if $P(XY) \neq P(X)P(Y)$. In this case, knowing whether X is present provides information about Y, and knowing whether Y is present provides information about X. Such nonindependence of DNA segments is known as *linkage disequilibrium*. In some cases, blocks of high linkage disequilibrium may be separated from each other by breaks at "hot spots" of recombination. But in many cases, linkage disequilibrium declines gradually with distance so that a chromosome cannot be divided into discrete evolutionary genes. Rather, there is a region of high linkage disequilibrium associated with each polymorphic site that can be considered the evolutionary gene with respect to that site, but evolutionary genes that are defined with respect to different polymorphic sites may overlap.

If X and Y are perfectly associated, then natural selection is indifferent to whether an effect (relative to non-X and non-Y) is due to X or Y alone, to the sum of their effects, or to their interaction. (The causal question can be addressed by experimentally breaking the association and producing X without Y or Y without X.) By contrast, if X and Y are randomly associated, then X can be treated as a variable part of the environment of Y and Y as a variable part of the environment of X. Clearly, the extremes of perfect and random association are the ends of a continuum. As linkage disequilibrium increases, it becomes more convenient to treat X and Y as parts of the same evolutionary gene. As linkage disequilibrium decreases, it becomes more convenient to treat X and Y as part of each other's environment.

Peter Godfrey-Smith (2009, 135–139) sees the inability to assign nonarbitrary boundaries to evolutionary genes as a major

flaw of gene-selectionism. In his view, "A Darwinian population is made up of a collection of definite countable things,"
but evolutionary genes are, at best, marginal Darwinian individuals because they fail to satisfy this "definite countable" criterion. He concedes that the lack of countability does not matter
much "if one's point of view is sufficiently pragmatic" and that
similar problems sometimes arise in counting organisms, cells,
and groups, but the difficulties are, in his view, particularly pronounced for evolutionary genes. They are not the sort of "real
entities that undergo the kind of change that Darwin described."
Rather than talk of genes, "in an evolutionary context it is more
accurate to talk of *genetic material*, which comes in smaller and
larger chunks, all of which may be passed on and which have
various causal roles."

Godfrey-Smith confounds two kinds of count in his calculation of the number of genes in a bacterial population. First, he
counts the number of genes in a bacterium (a few thousand).
Second, he counts the number of bacteria in the population (a
million). Then, he multiplies these numbers to obtain the number of genes in the population (a few billion). The first number is a count of different kinds of genes. This number is poorly
defined because boundaries between genes are indeterminate.
Moreover, the items so counted do not constitute a Darwinian
population. The second number is a count of how many genes
of each kind. This is the size of a Darwinian population, but the
count is not affected by where one places the boundaries between
genes.

Similar issues arise when counting evolutionary genes in
sexual eukaryotes. The number of genes on a chromosome is
poorly defined because boundaries between genes are fuzzy, but

the number of copies of the X chromosome in a group of organisms is not affected by where one places boundaries between genes. Only the latter number measures the size of a Darwinian population. If the population at a particular site consists of different variants ("alleles"), then selection can be measured by changes in the nonarbitrary numbers of these variants. Linkage disequilibrium is a measure of how representative the count at one polymorphic site is of counts at nearby polymorphic sites.

Many significant things do not have precise boundaries. No line on the ground demarcates where the Rocky Mountains begin and there is no way to count the number of peaks in Colorado without making pragmatic, somewhat arbitrary, choices about what counts as a peak. One could, if one chose, think of North America as made up of smaller and larger chunks of *landscape material* without naming any topographical features with fuzzy boundaries. Locations and areas could be identified by latitude and longitude, but this would be cumbersome. Evolutionary genes were conceptually central for Williams and Dawkins because such stretches of DNA persist over many generations whereas organisms, cells, and groups are ephemeral. Within their conceptual framework, discrete boundaries are unimportant but persistence is central. Within Godfrey-Smith's conceptual framework, Darwinian individuals should be clearly identifiable things, but persistence is not one of their essential properties.

DNA blocks defined by strong linkage disequilibrium need not respect the boundaries of protein-coding units. They may be smaller or bigger than such units. As George Williams noted, "Various kinds of suppression of recombination may cause a major chromosomal segment or even a whole chromosome to be transmitted entire for many generations in certain lines of descent. In such cases the segment or chromosome behaves in a way that approximates the population genetics of a single gene" (1966, 24). From this perspective, a mitochondrial genome or the nonrecombining portion of the Y chromosome can be considered a single evolutionary gene, as can the entire genome of an asexual organism.

The definition of linkage disequilibrium, $P(XY) \neq P(X)P(Y)$, can be generalized to all cases of nonindependence between X and Y. From this perspective, species boundaries are a major cause of linkage disequilibrium. For example, there is complete linkage disequilibrium between the genes of the native red squirrel and the introduced grey squirrel in the English countryside. Grey squirrel DNA has rapidly displaced red squirrel DNA from most British forests with the frequency of all parts of the red squirrel genome changing in concert relative to all parts of the grey squirrel genome. Some DNA segments from red squirrels might be selectively favored in grey squirrel bodies, but they never get the chance because red and grey squirrels cannot interbreed. Natural selection acts on phenotypic differences between the gene pools but does not "see" independent effects of smaller DNA segments. Ecological displacement can be considered a selective process in which the nonrecombining units are the gene pools of the competing species (G. C. Williams 1986). Proponents of the evolutionary gene concept could accommodate interspecific competition by identifying the gene pools as "evolutionary genes" or could avoid the issue by restricting the concept's application to natural selection within sexually recombining populations.

The Strategic Gene

What is the selfish gene? It is not just one single physical bit of DNA . . . it is all replicas of a particular bit of DNA distributed throughout the world.

—Richard Dawkins (1976)

A proper understanding of the units of selection problem must take account of an important symmetry: *Just as organisms are parts of groups, so genes are parts of organisms.*

—Elliott Sober and David Sloan Wilson (1994)

A gene that is distributed throughout the world cannot be part of an organism that is localized in space. Whether genes are the "unit of selection" has remained contentious, in part, because different meanings of "gene" are conflated. A first step to untangle this knot is to recognize that "gene" can refer both to a type and to tokens of the type (as well as to collections of tokens of a type). Gene tokens are physical objects but gene types are abstract kinds. It is tempting to simplify matters by suggesting that Dawkins refers to the type when he describes a gene as all replicas of a particular bit of DNA whereas Sober and Wilson refer to tokens when they identify genes as parts of organisms. However, such an attempt to cut the knot fails for the "selfish gene" because universal benevolence is not predicted when all members of a species possess tokens of the same type.

Evolution is often characterized as changes in gene frequency and the phenotypic effects of these changes. Changes in frequency imply counting, but gene tokens are rarely counted; rather, population geneticists usually lump together large numbers of tokens defined by the boundaries of individual organisms and count these *collections* as a single gene (Queller 2011). Thus, all tokens of a type in a haploid individual are counted as one gene (haploids have one allele), whereas all egg-derived tokens in a diploid individual are counted as one gene and all sperm-derived tokens as another gene (diploids have two alleles). This sleight of hand facilitates the development of simple mathematical models of evolutionary change.

Multilevel selection theory implicitly defines a gene as a single token within a cell in some contexts but as the collection of all tokens of a type within an organism in other contexts. Gene-selectionists also implicitly define the gene as a collection of tokens, but a collection that may be distributed across multiple organisms, for example across the members of a hive. I call a coterie of tokens that act together a *strategic gene* because it is this collection of gene tokens that can be considered a strategist in an evolutionary game played with other strategic genes.

A gene token is transcribed when an RNA polymerase copies its sequence into a functional RNA (such as a messenger RNA that is translated into a protein) and is replicated when a DNA polymerase takes apart the two strands of its double helix and uses each as a template to produce two new tokens. The strategic gene groups together tokens that cause an effect (actors) with tokens of the same type whose probability of replication is thereby affected (recipients). Actors may be located in somatic cells of a multicellular organism with recipients located in germ cells of the same organism, but actors and recipients may also be tokens of the same type located in different organisms. The strategic gene is not a fixed entity but can evolve to encompass more, or fewer, tokens of its type.

Consider a particular token in a germ cell (the focal token) and trace its ancestry back to the *urtoken*, the very first token of its type to arise by mutation. From the urtoken, a dichotomously branching tree can be envisaged that represents the history of all tokens of the type, with the focal token at one of the tips of the tree. The path through this tree from focal token to urtoken summarizes the selective history of the focal token (figure 4.1a). In organisms with a strong germ–soma distinction, the tokens on this path are located in germ cells, whereas most of the tokens on lateral branches are located in somatic cells. Tokens on the germ-path may be recipients of effects from tokens on lateral tips of the token-tree (figure 4.1b, c). Selection acts when these effects make a replicative difference (cause a change in relative frequency) relative to tokens of some other type. In this schema, phenotypic effects flow "inward" from somatic actors to germ-line recipients. These "causal arrows" influence which tokens are replicated but do not change the type of the token.

The extent of the strategic gene is determined by the number of replication cycles that separate the tokens responsible for a phenotypic effect from the tokens that thereby gain a selective advantage. Tokens on remote tips of a token-tree may be unable to exert selective effects on each other's replication because token-trees are broken up into selectively isolated fragments by spatial dispersal of tokens and mixing with tokens of other types. Selectively isolated tips of a token-tree belong to different strategic genes. Which tokens belong to a strategic gene is determined by the answer to the question, what is it about the effects of tokens of this type that accounts for a focal token being present in the population rather than a token of another type.

The strategic gene navigates a perilous path between the Scylla of the gene token (material gene) and the Charybdis of the gene type (informational gene). It is a collection of tokens but not the collection of all tokens of a type. Its tokens may be distributed across multiple levels of the hierarchy of interactors

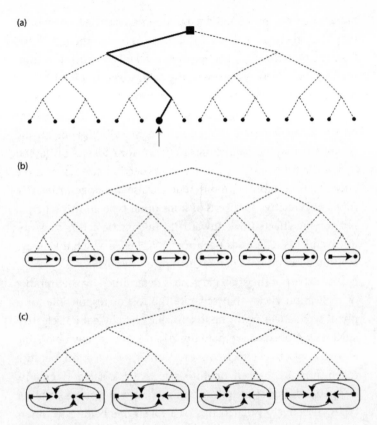

Figure 4.1
Representations of a token-tree: (a) Filled circles represent tokens of
a gene type. The arrow identifies the focal token. Its ancestry can be
traced back to the first token of its type (the urtoken represented by
a square). The token-tree represents the genealogical relationships of
all tokens descended from the urtoken. (b) A simple model in which
an actor token (square) confers a benefit (arrow) on a recipient token
(circle) of the same type. Groups of tokens that interact in this way correspond to strategic genes (enclosed in cartouche). (c) The strategic gene
can evolve to encompass more (or fewer) tokens of a type. Dichotomous
branching is a property of the semiconservative replication of DNA but
is not a necessary feature of the genealogies of genes. Rabies virus has a
single-stranded RNA genome. This strand is transcribed to produce the
complementary single-stranded antigenome that serves as a template to
transcribe multiple copies of the parent genome (Wunner 2007).

of multilevel selection theory, but it is not a level of this hierarchy. When viewed in this light, gene-selectionist and multilevel selection frameworks are fundamentally similar ways of describing the same phenomena. The strategic gene combines tokens responsible for a phenotypic effect with tokens of the same type that benefit, directly or indirectly, from the effect. It is the beneficiary of the effects it causes. As such, it is a *unit of adaptive innovation* and a unit of self-interest.

Historical Kinds

When I first used the type–token distinction to distinguish the informational gene (as type) from the material gene (as token), I thought of the informational gene as corresponding more or less to the classical notion of an eternal form. I now think it is more fruitful to think of the informational gene as a *historical kind* (Millikan 1999). It comes into existence, is copied, and ceases to exist when copying ceases or an imperfect copy becomes a new historical kind. Conceptualized in this way, the informational gene can be considered from either a diachronic (historical) or synchronic (contemporary) perspective. From the diachronic perspective, the informational gene is an individual with a distinct origin and history (Hull 1978), but from the synchronic perspective it is a historical kind, a collection of things that are similar because of common ancestry.

Strategic genes are also diachronic *individuals*, with a distinct origin and history. A strategic gene comes into existence when a token becomes causally "detached" from other tokens of its type, and consists of the descendants of this token until they lose their causal cohesion by "death" or "detachment." Although a strategic gene is an individual, it is not a synchronic historical kind because its tokens are indistinguishable from tokens of other strategic genes that together are the present parts of an informational gene.

The definition of the informational gene as a historical kind creates the seeming difficulty that if the same sequence originated independently in two lineages, there would be two informational genes that were indistinguishable. Would it not be simpler to consider the informational gene an eternal kind defined by an essence rather than its ancestry? My pragmatic answer is that the same sequence, if it has appreciable length, is highly unlikely to have independent evolutionary origins within the age of this universe. Of course, it is possible that two sequences that are already highly similar because of shared ancestry could converge on the same sequence, but there are pragmatic ways of dealing with such cases within the concept of a historical kind. A definition of the informational gene as an eternal chemical kind would create a peculiar ontology. Consider a DNA molecule of length 1,000 nucleotides composed of the four nucleotides A, T, C, and G. There are $4^{1,000}$ distinct eternal kinds of this length (for simplicity of exposition, I am ignoring the many molecular types that differ in ways other than nucleotide sequence). This number ($2^{2,000}$) vastly exceeds the number of elementary particles in our universe. Eternal genes are not parts of a practical ontology that carves nature at the joints because this would be an ontology with many more joints than things in the world.

Developmental Systems Framework

Developmental systems theory has been presented as a radical challenge to gene-centered accounts of development and, by extension, to gene-centered accounts of evolution (Gray 1992;

Griffiths 1998; Oyama 2000). In the developmental systems framework, genes are just one among many components of a developmental matrix and have no privileged causal role in development. Gene-selectionism is seen as misguided because it implicitly endorses a dualistic account of development in which genes are the carriers of preformed instructions of how to construct phenotypes, with the environment cast in a subsidiary and passive role.

Adaptation by natural selection takes a back seat to ontogenetic questions for proponents of developmental systems theory, whereas adaptation occupies the front seat for proponents of gene-selectionism. A simple resolution would be to propose that the two frameworks address different questions. Such a proposal would probably be perceived as partisan because the premise that ontogenetic and evolutionary questions require different kinds of answer is generally accepted by one side (Dawkins 1982, 98; G. C. Williams 1986) but rejected by the other (Gray 1992, 187; Oyama 2000, 45). Gene-selectionists believe that a conceptual separation of developmental from adaptive explanations aids clarity of thought, whereas developmental systemists believe such separation obscures more than it illuminates.

My intent has been a defense of gene-selectionism, not an attack on the fastness of developmental systems theory. Developmental systemists are on firm ground when they argue that genes do not have a privileged role in a causal account of development, and that phenotypes (in the traditional sense) are constructed by complex, highly nonadditive interactions of genetic and environmental factors (Gray 1992, 172–174). For these reasons, effects of individual genes on the course of development cannot be isolated in developmental time, but this is exactly what natural selection does over the course of many generations as genes are tested in different genetic backgrounds and in a series of environments. Natural selection extracts the average additive *effects* of genes as the environment chooses among phenotypes.

In the foreword to the second edition of Oyama's *Ontogeny of Information*, Lewontin remarked:

> Throughout the history of modern biology there has been a confusion between two basic questions about organisms: the problem of the origin of differences and the problem of the origin of state. At first sight these seem to be the same question, and taken in the right direction, they are. After all, if we could explain why each particular organism has its particular form, then we would have explained, *pari passu*, the differences between them. But the reverse is not true. A sufficient explanation of why two things are different may leave out everything needed to explain their nature. (2000, viii)[1]

By implication, understanding causes of difference is subsidiary to understanding causes of state. Lewontin (1974, 2000) and Oyama (2000, 52, 155) have perceptively identified an issue on which there is conceptual disagreement between the rival camps.

In the developmental systems framework, genes are nonprivileged components of a developmental matrix and it is a conceptual error to assign phenotypic features to genes rather than to the matrix as a whole. The entire matrix, or life cycle, constructs itself epigenetically in each generation. By the definitions of this chapter, a gene's environment is co-extensive with the developmental matrix except that the environment is defined

---

1. Isadore Nabi offered the rejoinder: "After all, if we could explain how each organism has evolved its particular form, by the selection of differences, then we would have explained, *per stirpes*, why it has its particular state. But the reverse is not true. A sufficient explanation of *how* an organism develops may tell us nothing about *why* it has its particular form."

to exclude the gene. There are effects of the gene, and all else is environment.

Effects are differences. Natural selection chooses *this* versus *that* based on a phenotypic difference. Why we observe these developmental systems and not others is explained, in part, by a long history of selection among differences and thereby selection of particular heritable difference-makers. Selection of differences can result in profound changes of state. One might say that there is selection *for* the causes of difference and selection *of* the causes of state.

Gene-selectionists use the language of statistics, of variances, correlations and average effects, whereas developmental systemists prefer "causal" accounts. The contrast, within physics, between mechanics and thermodynamics provides a useful analogy. Thermodynamics is a statistical theory, not an exact causal theory. It makes predictions that are right on average. In principle, a thermodynamic account of any system could always be superseded by a complete mechanical account, but in many circumstances an exact causal account is not practical, nor even possible, nor would it add much to the thermodynamic explanation.

There is, in principle although not in practice, a complete account of all evolutionary change expressed in terms of proximate physical causes that makes no appeal to concepts of selection, information, average effects, and the like. But I will settle for what is practical and predictive. Sober writes, "The strategy of averaging over contexts is the magic wand of genic selectionism. It is a universal tool, allowing *all* selection processes, regardless of their causal structure, to be represented at the level of the single gene" (1984, 311). I agree, but see this as a strength rather than a weakness of gene-selectionism.

A developmental system exists in which thornbill chicks in a nest are fed by a family of thornbills. This system is reconstructed in each generation as part of the life cycle of thornbills. The nest is a key developmental resource that is constructed afresh in each generation. There is another developmental system in which a cuckoo chick, in a similar nest, is fed by a family of thornbills until the cuckoo is larger than a thornbill. This system, including the nest, is reconstructed in each generation of the thornbill–cuckoo symbiosis. Developmental resources are similar in the two systems. The key difference-maker is the placement, in the nest, of a cuckoo egg with its cargo of cuckoo genes. Developmental systemists see these systems as fundamentally similar, whereas gene-selectionists see them as fundamentally different. Both are right. The origins of adaptation, by selection for the causes of average additive differences, and processes of development, involving highly nonadditive interactions among causes of state, are both fundamental questions.

Are Genes Special?

Many things besides genes are replicated, including membranes, song traditions, burrows, and nests (Sterelny, Smith, and Dickison 1996; Sterelny and Griffiths 1999, 70). Genes, however, possess a peculiar property that distinguishes them from most other heritable difference-makers. Hermann Muller observed that genes catalyze their own replication:

But the most remarkable feature of the situation is not this oft-noted autocatalytic action in itself—it is the fact that, when the structure of the gene becomes changed, through some "chance variation," the catalytic property of the gene may become correspondingly changed, in such a way as to leave it still autocatalytic. In other words, the change in gene structure—accidental though it was—has somehow resulted in a change of exactly appropriate nature in the catalytic reactions, so that the new reactions are now accurately adapted to produce more material just like that in the changed gene itself. (1922, 34)

This prescient passage was written before elucidation of the structure of DNA. We now understand, in considerable molecular detail, how this extraordinary property is achieved by each strand of the double helix acting as a template for the replication of the other strand. Not all chemical changes to DNA molecules are preserved through DNA replication, however (changes to the backbone are not maintained, but changes in the sequence of bases are). Moreover, mechanisms of proofreading and repair have evolved to correct "errors" of replication (Sterelny, Smith, and Dickison 1996). But some changes remain uncorrected, as are subsequent changes to these changes, allowing the exploration of a vast space of possible sequences. Muller recognized that this open-ended property of genetic change had far-reaching consequences:

Thus it is not inheritance and variation which bring about evolution, but the inheritance of variation, and this in turn is due to the general principle of gene construction which causes the persistence of autocatalysis despite the alteration in the structure of the gene itself. Given, now, any material or collection of materials having this one unusual characteristic, and evolution would automatically follow, for this material would, after a time, through the accumulation, competition and selective spreading of the self-propagated variations, come to differ from ordinary inorganic matter in innumerable respects, in addition to the original difference in its mode of catalysis. There would thus result a wide gap between this matter and other matter, which would keep growing wider, with the increased complexity, diversity and so-called "adaptation" of the selected mutable material. (1922, 35)

Not all things that are replicated have the property that changes in their structure, arising by chance and selected by the environment, are transmitted to future generations without compromising autocatalysis. And even if some nongenetic replicators transmit minor changes in this manner, few, if any, have potential for the open-ended adaptive change that is characteristic of DNA sequences. Human cultural evolution clearly has this open-ended quality (Boyd, Richerson, and Henrich 2011), although the nature of the heritable difference-makers of cultural change, if such exist, is disputed. One needs to have evolved very sophisticated organisms by other processes before meaningful cultural evolution can take place. Genes are special (and so, in its own way, is culture).

The strategic gene is a refinement of the metaphor of genes as self-interested agents. The phenotypes that are chosen by natural selection resemble those that would be chosen by a rational agent attempting to ensure its own transmission to future generations. Mindless genes can therefore be viewed *as if* they make strategic decisions. Some find this metaphor appealing (Dennett 2011; Queller 2011). Others consider it insidious and paranoid (Godfrey-Smith 2009, 144). Agential metaphors are less attractive (perhaps less seductive) for most other hereditary replicators, such as DNA methylation, membranes, nests, or money invested in the stock market. Are gene-selectionists inconsistent, or is there some principled difference between genes and nonagential replicators? I believe the difference resides in the sophistication of genes' strategies made possible by the open-ended nature of genetic inheritance identified by Muller. Genes are "indefinite hereditary replicators" (Maynard Smith and Szathmáry 1995, 58) that accumulate functional information about what works in the environment to a much greater extent than other candidates for the replicator role.

From *Whence Have We Come* to *Where Are We Going*

The formalism of the strategic gene redefines phenotype and environment. All choices involve selection among things that are different against a backdrop of things that are the same. In a "choice" of nature, that which is different is *phenotype* and that which is the same is *environment*. Evolution by natural selection is a process that converts phenotype, that which is selected, into environment, that which selects. It is a process by which large events, at the level of ecology and organismal behavior, determine fine structure, at the level of molecular gene sequences. It is the means by which the macrocosm shapes the microcosm.

The heritable difference-maker when two alleles differ by a single nucleotide is the nucleotide difference. This heritable difference is subject to selection on its phenotypic effects in the *environmental* context of all nucleotides that are the same. If one of the nucleotides becomes fixed in the population, then that nucleotide becomes part of the environment that selects among remaining differences. The "extended phenotype" and "niche construction" are sometimes presented as alternative labels for a single concept, but I would prefer a division of labor in which the former refers to differences under selection and the latter to evolved samenesses of the environment (Haig 2017). Bodies and genomes are constructed niches that select extended phenotypes of genetic differences.

Many readers will think that a "difference that makes a difference" is not what most people mean when they speak of a gene. That is true. In some contexts, it might be convenient to have another term for the concept of a heritable difference-maker. I have already suggested a term that fits the bill—"usit" as shorthand for unit-of-selection—but I am loath to introduce a new term because there is a long history of implicitly defining a gene as a difference, because "unit-of-selection" already has a tangled history, and because I have only created confusion with a previously attempted neologism that most readers loathe (see the Supplement to the Supplement to the Supplement to the Appendix to this book). But usit is now a historical kind. If people "use it," it will thrive.

Janus was the god of transitions. Natural selection involves a transition from a past difference to a present thing. Mutation is a transition from a past sameness to a present difference of possible futures. This chapter began by contrasting *genes as differences* and *genes as things*, but it is perhaps more fruitful to think of these as two faces of change. The selection of a gene, like the selection of a word, acquires meaning from what it is not. Each "choice" of nature requires a phenotypic *difference* but the outcome is a genotypic *thing* that embodies information about why it was chosen and that acts in the world. The next chapter considers genes as actors in present organisms and therefore focuses on genes as material things.

5   Limber Robots and Lumbering Genes

> Now [genes] swarm in huge colonies, safe inside gigantic lumbering robots, sealed off from the outside world, communicating with it by tortuous indirect routes, manipulating it by remote control.
> —Richard Dawkins (1976)

This famous sentence from *The Selfish Gene* is often interpreted to mean that organisms are puppets of their genes. Genes pull the strings. They are in control. At the same time, Dawkins subtly undercuts genetic autonomy. Genes are *remote* and *sealed off* from an outside world with which their communication is *tortuous* and *indirect*. Do genes make the decisive inputs to choice, or is control primarily an attribute of the robotic structures constructed by genes? Can the robot decide when to consult its genes? Can genes override robotic decisions? Are genes merely interested bystanders of a play performed for their benefit?

We construct robots to free ourselves from mundane decision making and to perform actions that we are unable to achieve without prosthetic aid. Some decisions in the control of robotic action are reserved for human actors but others are delegated to the robot. Consider a robot built to explore Mars. The robot interacts directly with the Martian environment, but its earthbound controllers experience Mars only vicariously through the robot's sensors. The robot and its controllers use inputs from these sensors to modify the robot's behavior. Not everything sensed by the robot need be communicated to Earth. The robot has other sensors that receive input from ground control. For some decisions, the robot is told what to do by its earthbound controllers; but for other decisions, it is on its own, because, at crucial moments, it must respond to events more rapidly than signals can be exchanged with Earth (Dennett 1984, 55).

One can imagine a spectrum of robot autonomy. At one end of the spectrum, robots are simple mechanical prostheses with all important decisions taken by humans. At the other end, robots are designed by humans but make all decisions on their own. Where do Dawkins's lumbering robots reside on this continuum? Are the robots on a short leash, with all important decisions taken by their genes, or are the robots fully autonomous, exploring and exploiting their environment without consulting their genes?

"Lumbering" has connotations of awkwardness, but some robots perform actions with a delicacy and precision that are beyond the powers of unaided human actors. The aim of robotic design is to produce supple rather than clumsy machines and the same is surely true of the "designs" of natural selection. An organism is an agile automaton (from the Greek, meaning "acting of itself") designed by natural selection to function effectively in a complex world. Simple automata (genes and proteins) interact with each other in complex networks to create a hierarchy of higher-level automata (cells, organs, organisms). Simple automata have only a small number of possible states, but as the number of simple automata that constitute a larger system increases, so does the number of possible states of the system, in a combinatorial explosion. As a result, higher-level automata can express more flexible behavior and possess more sophisticated information about their environment than can lower-level automata. Genes may be less nimble than the robots they construct.

An automaton "detects" a property of the environment when this property causes a change in the automaton's state. This environment may contain other automata. "Communication" takes place when one automaton (the sender) causes a change in state of another automaton (the receiver). A receiver detects a change of state of the sender, either directly, by physical contact, or indirectly, by the detection of a change in the environment induced by the sender's change of state. Genes and proteins can potentially fill the roles of sender and receiver, but so too can cells and other higher-level automata.

Genes encode instructions for the construction of protein automata but genes are themselves automata. A gene's states are determined by the binding of transcription factors and other proteins, by interactions with RNAs and with other DNA sequences, and by chemical modifications such as cytosine methylation. These factors determine where and when the gene is expressed. A gene may possess information about past environments as well as information about the current environment. An imprinted gene, for example, "remembers" whether it was present in a male or female body in the previous generation. A common criticism of the selfish-gene approach is that it assigns too much agency to genes. A gene is not a homunculus aware of everything that is happening to the organism and making plans accordingly, but one can also err on the side of underestimating the strategic options available to genes.

The principal way that a gene interacts with the world is by the production of RNA transcripts, some of which are translated into proteins. Many proteins have multiple states and function as simple automata. The factors that induce changes of a protein's state are what the protein "knows" about the world. A protein communicates with another protein when it induces a change in the other protein's state. Each protein's repertoire of functional states may include changes of conformation caused by interaction with things in its cellular or extracellular environment, as well as modifications of its chemical structure.

Most housekeeping functions of organisms are performed by proteins. Genes and proteins are equally mindless aperiodic polymers. Why should we privilege genes, rather than proteins, as the evolutionary actors on an ecological stage? We think differently about genes because genes are a very special kind of automaton. They are, to use Dawkins's term, *replicators*. Chemical changes to their structure are transmitted to their descendants, if descendants they have. By contrast, structural changes to proteins are not transmitted to their copies because proteins are not copied. Informational genes perform the role of evolutionary repositories of heritable information used to construct organisms. Although informational genes specify the *construction* of organisms, it does not necessarily follow that material genes *control* organisms.

Who Decides to Smile?

Rhodopsin is a gene expressed in rod cells of the retina but not in other cell types. Thus, *Rhodopsin* must, in some sense, sense when it resides in a rod cell and use this datum to switch between active and silent states. In its active state, *Rhodopsin* produces an mRNA that is translated at the endoplasmic reticulum to produce a protein that forms a covalent linkage with the chromophore 11-cis-retinal to form the visual pigment rhodopsin. The receipt of a photon of an appropriate wave-length causes isomerization of 11-cis-retinal to all-trans-retinal. This causes a conformational change in rhodopsin that is propagated to other proteins, triggering a complex biochemical cascade that culminates in hyperpolarization of the plasma membrane and discharge of a higher-level automaton, the rod cell (Okada et al. 2001; Ridge et al. 2003). *Rhodopsin* (the gene) encodes rhodopsin (the protein). The common use of the same name for gene and protein is an example of metonymy. (Geneticists' standard convention, used in this paragraph and the remainder of this chapter, is that genes are italicized whereas their protein products are not.)

Two important points can be made about this example. First, the photon is detected by the protein, not by the gene. The gene neither detects photons nor signals their presence. Rather, the gene directs the construction of protein automata that detect the presence of photons and signal their presence to other protein automata. Second, a *single* state of the gene produces an automaton that switches among *multiple* states in response to "environmental" cues. A gene can produce a protein with a "behavioral flexibility" that the gene itself lacks. There is no simple relation between the number of states of a gene and the number of states of the protein automata it constructs.

Consider a quintessential example of social communication in which rhodopsin plays a part. Multitudinous photons are received by the rod and cone cells of an infant's retina. The pattern of discharge of retinal cells initiates complex processes in the infant's brain that result in the recognition of the infant's mother and the coordination of a motor response: a smile. By an equally complex process, the baby's smile is detected by its mother and elicits a smile in return. The entire chain of events—from receipt of photons at the baby's retina to the contraction of muscles in its mother's face—takes place *without the causal intervention of changes in gene state.*

An exchange of smiles is possible because countless gene copies specify the production of innumerable protein automata in an untold number of higher-level automata (nerve cells and muscle fibers). These cellular automata are organized into two very high-level automata (mother and child) who are able to respond to each other's facial gestures. The limber robots communicate without consulting their lumbering genes. The development and maintenance of organism-level automata clearly involve coordinated changes in gene state but the exchange of smiles is too rapid for transcription and translation to play a role. Higher-level automata acquire and act on information that is unavailable to lower-level automata. No gene in the infant's genome perceives its mother's face.

A Warm Inner Glow

What is true of the interpretation of photons is also true of the interpretation of other sources of information about an organism's external and internal environment. Another mammalian example reaffirms the substantial autonomy of protein (and cellular) automata from direct genetic control. Information about external and internal temperature is detected by temperature-sensitive neurons, some located peripherally and others centrally. These inputs are integrated in the hypothalamus and other brain regions to coordinate thermoregulatory responses (Morrison 2004; Romanovsky 2007). One such response is the activation of nonshivering thermogenesis in brown adipose tissue. Uncoupling protein 1 (UCP1) resides in the inner membrane of the mitochondria of brown adipocytes. Activation of UCP1 causes a proton leak that uncouples mitochondrial respiration from oxidative phosphorylation. As a result, organic substrates are burned with the release of heat. The entire chain of events from stimulus (skin-cooling) to response (activation of nonshivering thermogenesis) may take place without the direct intervention of genes.

The efferent arm of this response (from brain to UCP1) involves a complex signaling cascade involving multiple protein automata. Brown adipose tissue is innervated by noradrenergic neurons of the sympathetic nervous system. When these neurons receive appropriate input from the brain, they release norepinephrine, which binds to β_3-adrenergic receptors (β_3ARs) at the cell surface of brown adipocytes, causing the release of a small protein (Gαs) within the cell. Gαs stimulates another protein, adenylyl cyclase, to produce cyclic adenosine monophosphate (cAMP). Via a further series of protein automata, increased cAMP causes activation of UCP1 in the inner mitochondrial membrane (Cannon and Nedergaard 2004; Nakamura and Morrison 2007; Romanovsky 2007).

Although changes of gene state do not play a direct role in the acute response to cold, such changes play important roles in modulating responses over longer timescales. The processes by which cold-exposure increases the thermogenic capacity of brown adipocytes are instructive. Noradrenergic signaling via β_3AR not only activates UCP1 (the protein) but also promotes transcription of *Ucp1* (the gene), promoting the translation of more UCP1 protein. Noradrenergic neurons also form synapses on preadipocytes that express β_1-adrenergic receptors. Cold-induced activation of β₁ARs promotes the differentiation of preadipocytes into mature brown adipocytes, a process that involves the transcription of many genes that are inactive in preadipocytes. As a result of these changes in gene expression, the inner mitochondrial membranes of brown adipocytes contain more copies of UCP1, and the total number of brown adipocytes is increased, in anticipation of the next cold exposure (Cannon and Nedergaard 2004).

This book is an attempt by one organism-level automaton to communicate to other organism-level automata, your good selves. Each of us has a set of genetic controllers "back on Earth" trying to pull the strings, but we make most decisions on our own. Our pains and our pleasures are the sticks and carrots that genes use to influence our decisions in pursuit of their ends. Genes do not care for us, know little about our world, and cannot agree among themselves. We should respect their suggestions, but not too much. If this book serves my genes' interests, it will be by a very indirect route. Our genes' purposes are not our own.

Mixed Messages

We think of machines as integrated wholes in which all parts work together to achieve common goals. One of the recurring themes of this book is that genes may have divergent ends even within an individual organism. Thus, the parts of the organismal "machine" may sometimes act at cross purposes in pursuit of conflicting agendas. This is not how we usually think about machines. Such internal conflicts suggest an alternative metaphor of a society of actors that must work together to achieve common goals but do not always act with unanimity. These alternative metaphors, of organisms as machines or societies, yield different insights into the basic processes within cells.

Cell biologists and behavioral ecologists both use the language of "communication" and "signals" but make different implicit assumptions about how signals evolve. Cell biologists are usually interested in signals that are transmitted within cells or between cells of a single body. Signaler and receiver are implicitly assumed to have identical interests. The question of whether signals are credible does not arise because signalers do not have incentives to deceive. In behavioral ecology, on the other hand, signaler and receiver are different individuals, potentially with conflicting interests. Receivers must decide whether signals can be trusted. Although behavioral ecologists recognize the possibility of conflicts between individuals, they usually assume that individuals have well-defined, unitary interests.

These two areas of inquiry are, of course, intimately linked. A behavioral signal is usually the external output of a complicated process of signaling among and within cells of the sender. The perception and interpretation of the signal usually involves an equally complex process of signaling among and within cells of the receiver. Despite these intimate connections, different kinds of questions are typically asked about signaling within individuals and signaling between individuals. Communication within individuals is usually viewed as a problem in signal engineering within the paradigm of the organism as machine. Relevant questions are how to send signals efficiently, how to cope with noise and interference from other signaling pathways, and how to correct errors. Questions of signal efficacy also arise for communication between individuals, but behavioral ecologists usually focus on question of signal credibility. Can a signal be trusted? What is the sender's motive? Does the sender have something to hide? Communication is seen as a social rather than a mechanical activity.

Neither cell biologists nor behavioral ecologists have given much thought to the implications of conflicts within the genomes of individual organisms. Current theory does not predict how an organism should behave if its responses are influenced by genes with conflicting interests. I suspect there is no general answer to this question and that answers for specific cases will require detailed knowledge of intricate molecular processes within cells. The genome has aspects of a fractious and poorly informed committee attempting to set policy, but with most decisions on how to implement policy taken elsewhere. This perspective raises new and unsettling questions. Is deception possible within an individual? Could different parts of an individual disagree over whether to send a signal to another individual? Would such a signal be sent?

Genomic Imprinting and Kinship

There are many sources of intragenomic conflict, but I will focus on antagonism between genes of maternal and paternal origin. This conflict can be illustrated with a minor modification of a famous thought experiment. J. B. S. Haldane wrote:

Let us suppose that you carry a rare gene which affects your behaviour so that you jump into a flooded river and save a child, but you have one chance in ten of being drowned, while I do not possess the gene, and stand on the bank and watch the child drown. If the child is your own child or your brother or sister, there is an even chance that the child will also have this gene, so five such genes will be saved in children for one lost in an adult. If you save a grandchild or nephew the advantage is only two and a half to one. If you only save a first cousin, the effect is very slight. If you try to save your first cousin once removed the population is more likely to lose this valuable gene than to gain it. (1955, 44)

Haldane's logic is simple. Your first-degree relatives, including your children and full siblings, have one chance in two of carrying a copy of a rare gene present in your body. Therefore, your genes would increase their number of surviving copies if you were to sacrifice your life for more than two first-degree relatives. A second-degree relative, such as a grandchild or nephew, has one chance in four of carrying a copy of the gene. You need to rescue five second-degree relatives at the cost of your own life to increase the number of surviving copies of your genes. A third-degree relative, such as a cousin, has one chance in eight, and a fourth-degree relative only one chance in sixteen. You need to rescue nine third-degree relatives or seventeen third-degree relatives to be marginally ahead in the genetic accounting. This calculus was formalized in W. D. Hamilton's (1964) theory of inclusive fitness in which an actor values the fitness of other individuals in proportion to their probability of sharing a randomly chosen gene of the actor.

Genomic imprinting refers to molecular modifications of genes, in either the mother's germline or father's germline, that provide a historical record (in offspring) of whether a gene resided in a male or female body in the previous generation. An intragenomic conflict lay hidden in Hamilton's theory because imprinted genes have effects that differ depending on the gene's parental origin. Consider the question whether your genes would be prepared to sacrifice your life to save three drowning maternal half-siblings. Haldane and Hamilton would probably have answered no, because they assumed that a gene's effects were independent of its parental origin and a randomly chosen gene in you would have one chance in four of being present in each half-sibling. But this answer averages a one-in-two chance that one of your maternal genes has copies present in each half-sibling and a zero chance that one of your paternal genes has copies in your maternal half-siblings. Your maternal genes would gain a substantial benefit from the nepotistic sacrifice of your life for three first-degree relatives, but your paternal genes would suffer a major cost from the rescue of nonrelatives.

In the presence of genomic imprinting, genes possess information about their parental origin and can employ conditional strategies to do one thing when inherited from a mother and something else when inherited from a father (Haig 1997). This means that internal genetic conflicts are possible in many of our interactions with kin because most relatives have different probabilities of carrying copies of genes we inherited from our mother and our father. The major exceptions are our descendants, who have equal probabilities of inheriting the genes we inherited from our mothers and fathers. Full siblings are another category of kin with equal probabilities of carrying copies of our maternal and paternal genes. However, in our evolutionary past the category of "siblings" would have contained a mixture of full and half-siblings. Therefore, sibling interactions are also predicted to be subject to some degree of internal genetic conflict.

Your feelings toward your children and grandchildren should not be associated with major internal conflicts, because your descendants have equal probabilities of inheriting the copy of a gene you inherited from your mother or the copy you inherited from your father. But the reverse is not true: your genes of maternal origin necessarily have copies in your mother and your genes of paternal origin are absent from your mother. Therefore, strong internal conflicts are predicted in an offspring's relations to its mother. Readers who are parents might reflect on the different flavor of their relations to their own parents (predicted to be associated with internal conflicts) and their relations to their own children (predicted to lack major internal conflicts).

Resolution of Intragenomic Conflicts

Knowledge of proximate mechanisms is required to understand how internal genetic conflicts will be resolved. The simplest resolution of the internal conflict between genes of maternal and paternal origin occurs if the relevant genes are unimprinted and therefore lack information about their parental origin. In the absence of imprinting, a gene is constrained to exhibit the same behavior when it is transmitted via eggs or sperm. Consider an imprinted gene for which natural selection has favored a higher level of gene product when the gene is maternally derived than when it is paternally derived. A "resolution" of this conflict occurs when the gene is silent when paternally derived but expressed at the maternal optimum when maternally derived. The nature of the resolution is reversed at a locus where natural selection favors higher expression when a gene is paternally derived. For such a gene, the unbeatable strategy is to be silent when maternally derived but produce the paternal optimum when paternally derived. I have called this the *loudest-voice-prevails* principle (Haig 1997).

The loudest-voice-prevails is a simple form of "conflict resolution." Whichever allele favors more product produces that amount, and the other allele produces none. Silencing of one of the two alleles at a diploid locus has a number of important consequences, of which I will mention two. First, alternative alleles at the locus have phenotypic effects when inherited from one sex but are without effect when inherited from the other sex. Therefore, alleles at a maternally silent locus will be selected solely for their effects on patrilineal fitness, whereas alleles at a paternally silent locus will be selected solely for their effects on matrilineal fitness (Haig 1997, 2000). Second, the loudest-voice-prevails principle reveals a sender's identity to the recipient. If both alleles are transcribed, a recipient of a signal (gene product) has no way of telling whether the sender is a maternal or paternal allele. If one of the two potential sources of a signal is reliably silent, then the signal reveals the identity of the signaler.

At a single locus, the loudest-voice-prevails principle suggests that whichever allele favors the larger amount "wins." However, most organismic outcomes are influenced by many genes. For example, maternal and paternal alleles may disagree over how much investment an offspring extracts from its mother. Paternal alleles at a demand-enhancing locus may produce their favored amount of a demand enhancer, but these effects may be countered by maternal alleles at another locus producing their favored amount of a demand inhibitor (Haig and Graham 1991). A "resolution" of this conflict is possible with maternal silencing of the demand enhancer and paternal silencing of the demand inhibitor (Haig and Wilkins 2000; Wilkins and Haig 2001). This resolution of the conflict has the form of a stalemate: the marginal cost of an increment of demand enhancer balances any benefit paternal alleles would gain from increased demand; likewise, the marginal cost of an increment of demand inhibitor balances any benefit maternal alleles would gain from reduced demand. In general, neither matrilineal nor patrilineal fitness is optimized at the stalemate.

An example will illustrate how such a stalemate can be expressed in a signaling system. *Insulin-like growth factor 2 (IGF2)* is a paternally expressed gene that promotes fetal growth. Its protein product, IGF2, binds to two receptors (IGF1R and IGF2R). IGF1R mediates the growth-promoting effects of IGF2. IGF2R is a decoy receptor (a deceptor) that binds IGF2 and transports it to lysosomes for degradation (Filson et al. 1993). In most eutherian mammals *IGF2R* is paternally-silent (Killian et al. 2001). Thus, a paternal gene produces a growth factor (IGF2) that is degraded by the product (IGF2R) of a maternal gene (Haig and Graham 1991). This can be considered a simple form of deception: *IGF2* sends a signal to IGF1R, but the message is intercepted by IGF2R before it reaches the intended recipient. In this example, there is no transfer of information between *IGF2* and *IGF2R* (the respective genes). Rather, the message from *IGF2* is intercepted by a protein produced by *IGF2R*.

The loci that influence an organismal outcome may have more than two sets of interests. In a 2006 study, I explored interactions among multiple "factions" with respect to a single organismal trait ("demand"). I found that the factions tended to align into two "parties": one favoring increased demand and the other favoring reduced demand. More theoretical work is needed to see whether this prediction can be generalized to conflicts over multiple traits.

A Cold Shoulder

The control of nonshivering thermogenesis is a potential arena of conflict between genes of maternal and paternal origin in species that huddle together for warmth. Heat generation by one individual reduces the heating costs of other individuals in its huddle and creates an evolutionary temptation of free-riding (not paying a fair share of the communal heating bill). If the members of a huddle include half-siblings, maternal and paternal alleles can disagree over how much heat to contribute to the common good. Specifically, when members of a multiple-paternity litter huddle together, paternal alleles are predicted to favor a lower set-point for the brown-adipose thermostat than that favored by maternal alleles (Haig 2004a, 2008a).

Genomic imprinting influences at least one step in the signaling pathway that activates nonshivering thermogenesis in brown adipocytes. Gαs is one of several protein products of the complex *GNAS* locus (Abramowitz et al. 2004). Both alleles of *GNAS* produce Gαs in most tissues of the body, but only the maternal allele produces Gαs in brown adipose tissue (Yu et al. 1998). A second gene product, XLαs ("extra large" αs), is produced by paternal *GNAS* and antagonizes the effects of Gαs in brown adipose tissue (Plagge et al. 2004). Gαs and XLαs mRNAs are transcribed from different *GNAS* promoters, and use alternative first exons, but share their remaining twelve exons. Thus, Gαs is produced by maternal *GNAS* and promotes nonshivering thermogenesis whereas XLαs is produced by paternal *GNAS* and inhibits nonshivering thermogenesis. Imprinted genes also influence the recruitment of extra heating units. Two paternally expressed genes, *Preadipocyte factor-1* and *Necdin*, produce proteins that inhibit the differentiation of preadipocytes into brown adipocytes (Tseng et al. 2005; Haig 2008a, 2010a).

Brown adipocytes are heat-generating automata. Their level of heat production is determined by the combined effects of unimprinted genes, maternally expressed imprinted genes, and paternally expressed imprinted genes. The loudest-voice-prevails principle predicts that imprinted genes that increase heat production in multiple paternity huddles will be maternally expressed whereas imprinted genes that reduce heat production will be paternally expressed. However, current theory has little to say about why some genes in a pathway are imprinted whereas others are not. Why is *GNAS* is imprinted in brown adipocytes but *UCP1* is not?

Imprinting can only make a phenotypic difference at loci for which gene dosage matters. If one active allele is as good as two, then silencing one allele makes no selective difference. Some effects of Gαs are dosage sensitive. For example, loss of one functional copy of Gαs causes osteodystrophy despite the expression of Gαs transcripts from both alleles in bone (Mantovani et al. 2004). The effects of Gαs may be particularly dosage-sensitive because many G protein–coupled receptors activate multiple signaling pathways via alternative G proteins with different α subunits. Thus, the precise stoichiometry of α subunits may determine the balance of signaling among pathways and the nature of the cellular response. For example, the β3AR of brown adipocytes signals via both Gαs and Gαi (Chaudhry et al. 1994). However, it seems unlikely that Gαs is the only dosage-sensitive step in the pathway from the detection of norepinephrine to the activation of nonshivering thermogenesis. Should a brown adipocyte generate heat only when heat serves the interests of maternal genes, of paternal genes, of unimprinted genes, or of something else?

Prader-Willi and Angelman Syndromes

> The paucity of spontaneous movement in infants with Prader-Willi syndrome, and their described placid nature, may result in decreased interaction with care-givers.
> —S. B. Cassidy (1988)

Deletion of the paternal copy of a cluster of imprinted genes at human chromosome 15q11–q13 causes Prader-Willi syndrome, whereas deletion of the maternal copy causes Angelman syndrome. Thus, Prader-Willi syndrome is caused by an absence of expression of paternal genes and Angelman syndrome by an absence of expression of maternal genes. Therefore, the former syndrome is predicted to exaggerate behaviors that benefit the mother at a cost to the child, and the latter syndrome to exaggerate behaviors that benefit the child at a cost to the mother (Haig and Wharton 2003). These syndromes' complex phenotypes suggest aspects of development and behavior that are sources of contention between genes of maternal and paternal origin (Holm et al. 1993; C. A. Williams et al. 2005). Together, these syndromes provide clues about a struggle for control that takes place in typically developing children who inherit copies of the imprinted gene cluster from both parents and whose behavior is determined by the balance of effects of paternally expressed and maternally expressed genes.

Infants with Prader-Willi syndrome are disinterested in feeding, suck very poorly, and are often fed directly through a tube to the stomach to ensure adequate nutrition (Cassidy 1988). Their cry is described as feeble, weak, squeaky, peculiar, or not sustained (Aughton and Cassidy 1990; Butler 1990; Miller, Riley, and Shevell 1999; Õiglane-Shlik et al. 2006). Infants with Angelman syndrome, on the other hand, obtain adequate nutrition and do not require tube feeding. Another contrast is found in sleep patterns. Infants with Prader-Willi syndrome exhibit excessive sleepiness, whereas infants with Angelman syndrome exhibit excessive wakefulness. This suggests that sleep in typically developing infants is disordered because the child is engaged in an internal struggle between genes of maternal origin that are endeavoring to put the child to sleep and genes of paternal origin that are endeavoring to keep it awake. An evolutionary interpretation is that paternally expressed genes (absent in Prader-Willi syndrome) have been selected to favor intense suckling and frequent waking to prolong maternal infertility after birth and delay the arrival of a younger sibling who would compete with the older child for the mother's care and attention. By contrast, maternally expressed genes (absent in Angelman syndrome) have been selected to favor less frequent waking and earlier weaning. When new mothers complain of exhaustion, their fatigue can be considered an adaptation (or extended phenotype) of genes that their baby inherited from its father (Haig 2014; Kotler and Haig 2018).

Angelman syndrome is characterized by positive affect and a smiling demeanor with frequent laughter (Horsler and Oliver 2006a). This contrasts with the negative affect of Prader-Willi syndrome (Isles, Davies, and Wilkinson 2006). Laughter in Angelman syndrome has been described as inappropriate and unprovoked, but careful behavioral studies suggest laughter is rare in nonsocial contexts and is particularly pronounced after eye contact (Oliver, Demetriades, and Hall 2002; Horsler and Oliver 2006b). In a comparison of thirteen children with Angelman syndrome with a matched group of children with other forms of intellectual disability, the children with Angelman syndrome smiled more, were more likely to reach toward or touch adults before smiling, and their smiles were more effective at eliciting adult smiles in return (Oliver et al. 2007). The frequency of laughing and smiling decreases as children with Angelman syndrome grow older (Adams, Horsler, and Oliver 2011). Thus, young children with Angelman syndrome have been proposed to exhibit exaggeration of behaviors, laughing and smiling, that normally function to elicit maternal care, attention, and attachment (Brown and Consedine 2004; Isles, Davies, and Wilkinson

2006). Past social environments have evolutionarily shaped responses to smiles and have thereby shaped communication within cells.

The exuberant personality of children with Angelman syndrome is combined with profound deficits in verbal and nonverbal communication. Infants possess an abnormal high-pitched cry (Clayton Smith 1993) and babbling is delayed or absent (Yamada and Volpe 1990; Penner et al. 1993). Nonverbal communication is primarily used for making requests and rejecting offers (Didden et al. 2004). Such communication usually involves direct manipulation of the other person—pushing a hand away, leading by the hand, touching to gain attention—rather than gesture or pointing (Jolleff and Ryan 1993). Joint attention, joint action, and taking turns are poorly developed (Penner et al. 1993). Affected children never learn to speak (Clarke and Marston 2000; C. A. Williams et al. 1995). The absence of speech appears out of proportion to the underlying level of cognitive impairment (Alvares & Downing 1998; Pembrey 1996; Penner et al. 1993). Children with Angelman syndrome have poor motor imitation skills, and most fail to imitate verbal behavior (Didden et al. 2004; Duker, van Driel, and van de Bercken 2002; Jolleff and Ryan 1993; Penner et al. 1993). "Motor theories" of the evolution of human language posit that language is based on the perception and imitation of gestures of the vocal tract (Galantucci, Fowler, and Turvey 2006; Gentilucci and Corballis 2006). An interesting hypothesis is that the ataxia and absence of speech in Angelman syndrome have a common etiology in defects in the neural representation of motor actions.

The absence of speech with the absence of expression of maternally derived genes is intriguing. Badcock and Crespi

(2006) have suggested that genes of maternal origin have been selected to act in the language centers of the child's brain to promote attentiveness to maternal instruction and maternal example, coordinating maternal and child needs for the benefit of the matriline. Verbal communication appears to be arrested at a very early stage of development (Grieco et al. 2018). Perhaps earlier onset of language in children reduced costs to mothers and the genes whose expression is disrupted in Angelman syndrome normally function to initiate language development.

E pluribus unum

Let us adopt, for a moment, the perspective of developmental systems theory, and turn again to the question whether genes control organismal automata. Ontogeny involves necessary interactions between the material parts of the developing organism and its environment. Genes are important parts of this process but cannot act on their own. Nucleic acids, the material stuff of genes, are molecular components of the organism, no different from proteins, fats, carbohydrates, and minerals. These molecular components are organized into higher-level components— muscles, nerves, bones, glands—that work together to act purposefully in the environment. Changes of gene expression play little role in the moment-by-moment control of organismal behavior. At longer timescales of development, genes are tools used to remodel the organism in response to environmental inputs. The sense of gene invoked in this account of organismal function is close to the gene token or material gene. Material genes do not control organismal behavior. The organism, in its manifest complexity, controls itself. This is the metaphor of the autonomous robot.

How should we think about control if the whole can be divided against itself? The metaphor of a society suggests more flexible ways of thinking about agency. Consider the agency of nations and their citizens. Nations act in the world, declare war, sign treaties, invest in infrastructure, resolve conflicts among and discipline their citizens. These actions of nations are partially determined by actions of citizens whose choices and preferences are shaped by actions of nations. One can believe in the independent agency of citizens without denying the collective agency of nations. But the problem of understanding the causal relations between societies and their members is not simply the hermeneutic circle of making sense of the whole by reference to the parts and making sense of the parts from their place in the whole. Human individuals are not simply subsidiary parts of social groups. Groups may have overlapping memberships. Some individuals, for example, may be citizens of more than one nation. Dual allegiance can be a source of conflict within nations and a facilitator of cooperation between nations.

I recognize two kinds of actors on the timescale of organismal behavior: the historical individuals we identify as organisms and the historical individuals I have called strategic genes. Their relations resemble, in some respects, the relations between nations and citizens. Organisms act in the world in ways that are determined by the aggregate actions of their strategic genes, but individual genetic actions are determined by information processing at the level of the organism. Strategic genes are not merely parts of an organismal machine, because coteries of strategic genes within organisms may subvert the good of the whole for partisan advantage, as may coteries of strategic genes distributed across organisms. Strategic genes and organisms have different kinds of agency. They are different ways of carving nature at the joints.

Let us now turn our attention from the synchronic axis of development to the diachronic axis of evolution. On an evolutionary timescale, *informational genes* are texts. They are repositories of information about what has worked in past environments and what is anticipated to work in future environments. These texts have been written and revised by past environments and are inscribed in the medium of nucleic acid sequences. They include specifications for the construction of organisms that are present-time interpreters of the environment and readers of the genetic text. The present environment provides the context for interpretation of the material text over the course of present development.

6 Intrapersonal Conflict

For the good that I would I do not: but the evil which I would not, that I do.

—Romans 7.19, King James Version

In the chapter on will from his *Principles of Psychology*, William James discussed five types of decisions. Most decisions he noted were decisions without effort, but in the

final type of decision, the feeling that the evidence is all in, and that reason has balanced the books, may be either present or absent. But in either case we feel, in deciding, as if we ourselves by our own willful act inclined the beam: . . . If examined closely, its chief difference from the former cases appears to be that in those cases the mind at the moment of deciding on the triumphant alternative dropped the other one wholly or nearly out of sight, whereas here both alternatives are steadily held in view, and in the very act of murdering the vanquished possibility the chooser realizes how much in that instant he is making himself lose. It is deliberately driving a thorn into one's flesh; and the sense of *inward effort* with which the act is accompanied is an element which sets the fifth type of decision in strong contrast with the previous four varieties, and makes of it an altogether peculiar sort of mental phenomenon. (1890/1983, 1141)

After consideration of the kinds of decisions that were made with and without effort, James concluded that "effort complicates volition . . . whenever a rarer and more ideal impulse is called upon to neutralize others of a more instinctive and habitual kind."

Religious, literary, and psychoanalytic texts abound with discussions of conflicts between our higher and lower natures, between passion and reason, between selfishness and selflessness, between immediate gratification and pursuit of long-term goals. We all are familiar with being caught on the horns of a dilemma, of wanting to make a phone call and simultaneously not wanting to make the call, of being torn between temptation and conscience; but evolutionary biology has had little to say on why our subjective experience should be organized in this manner. At first sight, the idea that we can be at war with ourselves appears paradoxical. If we are products of natural selection, superbly designed to maximize inclusive fitness, why do we often find it hard to make decisions and stick to them? A fitness-maximizing computer would simply calculate the expected utilities of the different alternatives and then choose the alternative with the highest motivational score. Why should some kinds of decisions be more difficult to make than others? Is the subjective experience of effort merely a measure of the computational complexity of a problem, or is something else going on?

For William James, "The existence of the effort as a phenomenal fact in our consciousness cannot of course be doubted or denied. Its significance, on the other hand, is a matter about which the gravest difference of opinion prevails. Questions as momentous as that of the very existence of spiritual causality, as vast as that of universal predestination or free-will, depend on its interpretation" (1890/1983, 535). My aim in this chapter is not to address such momentous questions, nor to shed light on really difficult questions such as how and why we have subjective experiences. Rather, it is to ask how one might begin to reconcile nonbiologists' perception of the ubiquity of internal conflict with biologists' view of the mind as an adaptive product of natural selection. Internal conflict often *seems* maladaptive; consuming time, energy, and repose. If so, why does it persist?

Three kinds of hypotheses could resolve the conundrum of conflict within an adapted mind. First, one might argue that internal conflict arises from constraints on the perfection of adaptation; that evolved mechanisms work well on average but occasionally malfunction. We would be better off without internal conflict, but we are stuck with it. Second, one might argue that internal conflict is adaptive and the "contending parties" have the same ultimate ends and their conflict is in a sense "illusory." Natural selection has simply adopted an adversarial system as the best mechanism of arriving at useful truths. Finally, one might argue that internal conflict is "real" and reflects a disagreement over ultimate ends among multiple agents that contribute to mental activity. I will reveal my hand at the outset. I believe that all three kinds of explanations, together with their complex interactions, contribute to experiences of internal conflict.

First, let me briefly consider nonadaptive interpretations of intrapersonal conflicts. The precision of achievable adaptation is limited because natural selection is retrospective, fitting us to the past rather than the present environment; because the adaptive response to environmental change is limited by the pool of available variation and by time-lags until the origin of appropriate new mutations; and because selection is blind to very weak selective forces (i.e., chance plays a large role in who survives and reproduces when differences in adaptedness are slight). Some internal conflicts may merely reflect the imprecision of adaptation. Our genomes evolve by minor revisions to an old text just as the operating systems of computers evolve by the addition of new functionalities to old code; neither programmers nor natural selection have been able to eliminate all opportunities for malfunction. On this view, some internal conflicts may be analogous to "system conflicts" that occasionally cause my computer to crash: multiple functional programs are running simultaneously and occasionally make contradictory or ambiguous demands on the operating system. The analogy is, of course, limited. My computer does not, in fact, run multiple programs simultaneously. Instead, it is a serial machine that has only a single program running in its central processor at any particular moment, but switches rapidly between programs. Our brains, by contrast, are massively parallel processors with different subsystems handling different kinds of data. Somehow this dispersed neural activity has to be integrated in coming to a decision. Perhaps "conflict" could arise from imperfections in the process of integration.

Without doubt, our current environment presents us with novel challenges for which we lack specific adaptations. There were no opportunities in our evolutionary past to put aside resources for ten or twenty years, and then recover them with interest. Retirement planning is a recent cultural innovation for which we are unlikely to have evolved dedicated mechanisms. Instead, we employ general-purpose problem-solving machinery to make plans that come into conflict with more hardwired responses. My rational resolution to save is thwarted by short-term impulses that fritter away income on ephemeral goods. (In this case, it is unclear that a comfortable retirement has anything to do with enhancing fitness. From a genetic perspective, our impulses may have it right.) Powerful narcotics are another novelty for which we are adaptively unprepared. An addict may strongly desire to be free of his compulsion, but may lack the will to override strongly maladaptive, albeit evolutionarily programmed, cravings.

Adaptive explanations of internal conflict often invoke competition among alternatives as the most effective mechanism of choosing the best course of action. Consider a gazelle who is racing toward a stump before which he must either zig to the right or zag to the left. As the decision point approaches, the gazelle calculates probable outcomes if he zigs or he zags, in response to new information about the terrain to either side and the fear from which he flees. These plans of action compete against each other until he abruptly commits to one of the alternatives. The cheetah, in hot pursuit, also needs to plan ahead for both eventualities, to respond rapidly once the gazelle commits but not be fooled by a feint to one side. There is a premium on minimizing reactions times. The gazelle commits to one of his options and the cheetah responds with one of hers. The gazelle's options for zigging or zagging are in "conflict" until action, but "at the moment of deciding" the rejected alternative is dropped wholly from view. Such a model seems inadequate to explain "driving a thorn into one's flesh."

In an earlier chapter on instinct, James had considered conflicts of impulses that arose in one of two ways. First, an individual might have instinctive impulses to respond to O with A and P with B, but O could become a *sign* of P *from experience*, "so that when he meets O the immediate impulse A and the remote impulse B struggle in his breast for the mastery" (1011). Second, "*Nature implants contrary impulses to act on many classes of things*, and leaves it to slight alterations in the conditions of the individual case to decide which impulse shall carry the day" (1013). In both these models, James envisioned *experience* as deciding between contradictory impulses. The adjudication could be partly mediated by reason: "Reason, *per se*, can inhibit no impulses; the only thing that can neutralize an impulse is an impulse the other way. Reason may, however, *make an inference which will excite the imagination so as to let loose* the impulse the other way" (1013). This model of conflicting impulses resolved by experience can be likened to a court with a disinterested judge but, although James writes of a struggle for mastery between impulses, it is not immediately clear why such a process should be accompanied by feelings of effort.

Humans have evolved general-purpose problem-solving mechanisms (reason), and the ability to learn from others, to compensate for the limitations of hardwired instincts. We are rational, cultural, and instinctive beings. Sometimes these alternative sources of behavioral guidance promote different choices. Instinct summarizes the wisdom of past natural selection and recommends actions that have worked before under similar circumstances. Culture also summarizes wisdom from the past and can respond much faster than gene sequence to environmental change, but, from a gene's-eye view, has the disadvantage of evolving by rules that need not promote genetic fitness. Reason can respond to unique features of the current situation, and to weak selective forces, but may lack the historical judgment of either instinct or culture. Our passions, both positive and negative, are the carrots and sticks employed by genes to mold our actions to their ends. Reason may be a slave to the passions, but reason pursues pleasures as ends in themselves rather than as means to an end. (Coitus with rubber is an obvious example of reason circumventing genes' ends.)

Our ability to reason is an adaptive response to the imperfections of instinct, but this adaptation must have its own imperfections, including inevitable clashes when instinct and reason offer conflicting advice. But if such clashes are unavoidable and recurrent, humans should have evolved adaptive (although imperfect) mechanisms of resolving them. How might a well-designed organism resolve conflicts between the dictates of instinct and reason? Such an organism might have a limited ability to override instinct given strong enough reasons, with "strong enough" calibrated to match the strength of past selection favoring the instinctive response. Very strong motivation would be needed for reason to prevail in decisions closely related to fitness, for which instinct provides a powerful guide, but the threshold of motivation could be lowered when the prescripts of instinct are less strong. These considerations suggest an adaptive explanation for the feeling of effort in making certain kinds of decisions. Some decisions are hard to make because individuals in the past who made similar decisions with greater ease left fewer offspring. The strength of our will, no less than the strength of our muscles, can be shaped by natural selection. Moralists may derive some comfort from the muscular response to exercise.

Reality is more complicated than the above naïve model. Instincts are not unitary; neither is reason (itself a special kind of instinct). Different parts of the brain undertake different tasks, and no part has access to the big picture. An arrangement in which mental modules compete for attention and influence could be a general organizing principle of the mind. Different modules process different kinds of data to produce recommendations for action. What a module communicates to the decision-making collective may be no more than its preference rather than a detailed justification. The preferences must then be aggregated to generate a choice. Kenneth Arrow (1963) proved that no procedure for aggregating preferences in situations of social choice can be guaranteed never to violate basic axioms of rationality. His proof assumed that information was limited to the rank-ordering of preferences, and specifically did not allow interpersonal comparisons of the strength of preferences. Are there similar constraints on the rationality of intrapersonal aggregation of preferences?

In my subjective experience, choice becomes more difficult when motivations are not expressed in a common currency. Life would be simpler if one could simply compare expected degrees of pleasure from giving way to a sexual infidelity and from resisting it. Instead, the two courses of action promise rewards that differ in kind. There may be functional reasons for having more than one currency of reward—some rewards could be more suited to sustaining long-term projects, others for providing immediate gratification—but their existence renders intrapersonal comparisons problematic. Multiple currencies would still allow facile comparisons among rewards if their exchange rates were well defined, but this does not seem to be the case. Why should this be so? If one part of my mind counsels one action and another part counsels another, how is this disagreement to be resolved if there is no common unit for comparison?

Perhaps there are benefits, as well as costs, to not expressing all values on a single scale. Multiple currencies allow exchange rates to vary. This might be adaptive if the optimal weights to be attached to the pleadings of different internal voices vary from place to place and time to time. A woman who yields to an extramarital passion may face very different consequences in New York and Riyadh. Multiple currencies could allow her to learn the appropriate exchange rates for her culture from her

experience of which choices were rewarded and which punished during her behavioral maturation. If so, choices might
be expected to become easier as we grow older and learn the
norms of our culture (assuming these norms are not rapidly
changing).

The discussion so far has concerned questions of adaptation and constraint: there may be adaptive reasons for allowing
internal voices to compete for attention, but some expressions of
this competition may be maladaptive. No mechanism is perfect.
Pathological indecision may be just that: pathological. A model
of the mind has been presented in which mental modules may
express different preferences because they have different capabilities and process different inputs. In this model, the problem of
preference aggregation reduces to a problem of determining the
best weights to assign to the preferences of different modules.
However, another possibility should be considered: internal factions may disagree about ultimate ends. Each faction may have
an incentive to overstate the case for its side of the argument,
even for broadcasting misleading information.

Agents with different interests may have different preferences despite access to the same information. If the self is an
assemblage of agents with distinct interests, then internal conflict may reflect disagreements over ultimate ends. What benefits one, need not benefit all. Factions may disagree about the
weights that should be assigned to the recommendations of different parliamentary committees. There need be no agreement
on exchange rates, even when cultural norms have been fully
internalized. In this view, decision making would resemble the
deliberations of a collective: sometimes consensus is achieved,
sometimes one set of interests overrules the others, and sometimes the committee fails to decide.

I have in mind two kinds of agents with stakes in the deliberations of self. The first are genes. Genes' ultimate ends are the propagation of their copies. Put another way, the genes that we see today are those that have successfully propagated in the past. Genes can be said to have *purposes* to the extent that they possess properties that have promoted their own survival and replication. The second are ideas (or, to use Richard Dawkins's term, memes). Ideas may be propagated from other minds or generated afresh within a mind. (In fact, most of our ideas are hybrids that recombine content acquired from other minds with features generated in our own mind.) The ideas that reach our awareness have succeeded in competition with other ideas for attention. The ideas that colonize other minds have succeeded in competition with other ideas for expression by the transmitter and in competition with other ideas for perception by the receiver. Ideas can be said to have *purposes* to the extent that they possess properties that have promoted their propagation from mind to mind. Such properties are the "adaptations" of ideas.

Ideas compete for space on the page. There has been conflict in my mind over the writing of this chapter. I have often been undecided about what to write and this indecision has preoccupied my mind, crowding out other concerns. Different ideas and forms of expression have competed for expression. Many sentences were written, only to be erased and replaced by another sentence that I had temporarily discarded. Slowly a final text took form. What were the properties that made an idea successful in this competition? One of them was coherence with the rest of the manuscript. Another was my estimate of how likely an idea was to attract your attention. I preferred pithy to dull formulations. Another criterion, I hope, was some degree of correspondence with reality, what one might call "truth." Now some of these ideas have entered your mind, gentle reader. May they go forth and multiply.

Why do I care whether my book is read? Why do people, in general, care about the propagation of their ideas? The fact that we care suggests there has been an evolutionary correlation between the transmission of ideas and of genes. In other words, successful propagators of ideas have, on average, also been successful propagators of genes. Ideas can be useful, helping oneself and one's kin to survive in a challenging physical environment. And ideas can function in display: I want to impress you with how smart I am; I do not want to say anything silly; I want my ideas to spread so that I can bask in their reflected glory. Being a propagator of successful ideas must, on average, have translated into influence and control of resources. We have an instinct to generate and propagate ideas, and this instinct creates the environment in which ideas can compete and evolve their own purposes.

"Good" ideas are persuasive. They appeal to our genetic biases, even if they do not promote genetic fitness. The correlation between genetic and memetic propagation cannot have been perfect, and, at times, ideas may have propagated at the expense of genes, or genes at the expense of ideas. When a person willingly dies for a political or religious ideal, the incitements of ideas have triumphed over the urgings of genes. But, when a charismatic preacher compromises his ministry for a brief sexual encounter, the entreaties of his genes have trumped those of ideas. Natural selection favors the evolution of genetic biases against adopting ideas that reduce genetic fitness, but ideas evolve much faster than genes and will accommodate themselves to these new biases. Our choices are shaped by the interplay of the stubborn intransigence of genes and the supple agility of ideas.

But surely, one might object, we should evolve a consistent set of genetic biases. Not necessarily, if we are also subject to conflict among genes. Biologists commonly assume that all genes of an individual have the same interests because all have the same chance of being transmitted to that individual's progeny. For most choices, this is probably a reasonable assumption. However, there are subtle ways in which genes can have distinct interests, and these can promote contradictory adaptations within the genome. Transposable elements replicate faster than the rest of the genome. Nuclear genes are transmitted via eggs and sperm; mitochondrial genes are transmitted only via eggs. If different genes have different rules of transmission, then an adaptation that promotes the long-term propagation of a gene may not promote the transmission of the other genes with which it is temporarily associated. In the remainder of this chapter, I will concentrate on the specific case of conflict between the nuclear genes we inherit from our fathers and those we inherit from our mothers, but I will first digress on how natural selection acts on interactions among relatives.

We all die. A gene in a neuron does not leave *direct* descendants. Only genes that are present in our sperm or eggs have any chance of producing copies that survive our death. Nevertheless, each gene in the neuron is a copy of a gene in the fertilized egg that gave rise to both our brains and our gonads, and each gene in the egg has left its copies in both organs. For this reason, genes in brains have evolved complex adaptations that promote the propagation of their *indirect* copies in gonads.

Our body contains multiple kinds of cells, all with the same set of genes but with different roles to play in facilitating the transmission of gene copies via our gonads. If a gene in my liver can promote the transmission of its indirect copies in my gonads, then there is no reason why it cannot also promote the transmission of its indirect copies in the gonads of my relatives. When a mother provides milk to her baby, the genes in her breast promote the propagation of gene copies in the baby's gonads, not the mother's gonads. Lactation is associated with temporary infertility of the mother, delaying the birth of her next child. Thus, nursing promotes the propagation of gene copies in the baby's gonads at the expense of gene copies in the mother's gonads. There is no guarantee that a gene in the mother's breast has indirect copies in a particular baby. Rather, a gene in her breast has one chance in two of having an indirect copy transmitted to the baby. (The mother receives half her genes from her mother and half from her father. She transmits half to her baby, but this half is a mixture of genes she received from her mother and father.)

Among cooperatively breeding meerkats (a kind of mongoose), daughters sometimes help their mothers by nursing younger siblings. By this means, genes in the daughter's breast may promote the propagation of their indirect copies in the mother's gonads, because the mother is relieved of some of the costs of lactation, and in the sibling's gonads, because the sibling is the direct beneficiary of extra milk. Each gene in a daughter's breast has one chance in two of having indirect copies in the gonads of a sibling receiving extra milk, the same as the probability that the gene would have indirect copies in the gonads of one of the daughter's own offspring. If genes in breasts gain equivalent benefits from providing milk to offspring or siblings, why is sororal suckling not more common among mammals?

The equivalence of these two routes of promoting genetic fitness is predicated on the assumption that donor and recipient are full siblings, sharing the same mother *and* father, as is usually the case in meerkats. Suppose, however, that the younger sibling has the same mother but a different father. A gene that a daughter inherits from her mother has one chance in two of having an indirect copy in the younger sibling and one chance in two of having an indirect copy in the daughter's own offspring. The two routes of achieving fitness remain equivalent for genes of maternal origin. However, a gene that the daughter inherits from her father is absent from the younger sibling but has one chance in two of having an indirect copy transmitted to the daughter's own offspring. The two routes are not equivalent for genes of paternal origin. That is, maternal genes of the daughter are "indifferent" as to whether an offspring or a maternal half-sibling receives a benefit, but the daughter's paternal genes would "prefer" her own offspring to receive a benefit instead of a maternal half-sibling. Thus, under some circumstances maternal and paternal genes in a daughter's breast could "disagree" over whether to supply milk to a half-sibling.

The above simple example illustrates the general point that most relatives, excepting our direct descendants and full siblings, are genetic kin of our mother *or* father but not of both. Genes of paternal origin should "care" about their effects on the father's side of the family, but not the mother's side, whereas the situation is reversed for genes of maternal origin. Genes of maternal and paternal origin thus occupy different "social environments" and may evolve different behaviors suited to their different circumstances. A biblical example can illustrate asymmetries of maternal and paternal kinship. Ishmael was Abraham's son by Hagar, an Egyptian slave. While Ishmael dwelled with Abraham's extended family he would have been surrounded by many individuals with indirect copies of the genes he received from Abraham, but no individuals with indirect copies of the genes he received from Hagar, excepting Hagar herself. If Ishmael were to have performed some action that betrayed his father's household but benefited himself, this action could well have had negative consequences for the genes he received from Abraham, because of costs to the patriarchal family, but positive consequences for the genes he received from Hagar, because of benefits for himself. If Ishmael faced such a choice and similar situations had been repeated many times during human ancestry, one might expect his maternal genes to add a little weight to the scales on the side of betrayal and his paternal genes to add a little weight on the side of faith in the patriarch.

Asymmetries of relatedness are greatest in an offspring's relations to its parents, and it is here that the potential for internal genetic conflict is predicted to be strongest. A daughter's maternal genes are definitely present in her mother, but the daughter's paternal genes are absent from her mother. Therefore, the maternal genes of daughters value benefits to mothers as highly as benefits to self, but benefits to mothers are of little if any value for the paternal genes of daughters. (Here, I gloss over some complexities due to variation in mating systems.) Significantly, although internal genetic conflict is present in a daughter with respect to her relations with her mother, the same is not true of the mother's relations with her daughter. The mother's maternal and paternal genes have equal probability of being present in the daughter. Thus, a child's feelings toward its parents are predicted to be more internally conflicted than the parent's feelings toward the child.

Consideration of conflict between maternal and paternal genes would be mere sophistry if not for the fact that some genes exhibit behaviors that vary with the sex of their most recent origin. This phenomenon is known as *genomic imprinting* because it is assumed that some mark, an imprint, becomes associated with a gene in a parent's gonad. This imprint must then be transmitted to the gene's direct copies in the next generation, identifying the gene as coming from a mother or father, but the imprint must be able to be erased in the gonads of offspring, so that the gene's copies can inherit the appropriate imprint in grand-offspring. An imprint is a contingent, rather than a fixed, property of a gene because a paternal gene in a daughter will be a maternal gene in the daughter's son. Therefore, the paternal imprint must be erased and reinscribed as a maternal imprint. Imprinted genes are actively expressed when inherited from one sex, but silent when inherited from the other. Thus, a past environment, whether a gene resided in a male or female body in the previous generation, can influence the gene's expression in the current generation.

Imprinted genes influence brain development and function. Barry Keverne and colleagues were able to produce mice with two different kinds of cells in their bodies (Keverne et al. 1996). Some cells possessed genes inherited from a mother and father, whereas other cells possessed genes inherited from only one sex. Cells that lacked maternal genes were well represented in the hypothalamus of the mouse brain but absent from the neocortex, whereas cells that lacked paternal genes were well represented in the neocortex but absent from the hypothalamus. These observations suggest that maternal and paternal genes perform different roles during normal development of the mouse brain, and hint that paternal genes favor relatively greater weight for hypothalamic preferences in decision making whereas maternal genes favor relatively greater weight for neocortical preferences.

As a gross oversimplification, the hypothalamus controls "visceral," and the neocortex "cerebral," motivations. Among primates, the relative sizes of neocortex and hypothalamus correlate with the composition of social groups; in particular, neocortical size increases with the number of adult females in a typical group (Keverne et al. 1996). Two asymmetries of mammalian social groups may explain a bias toward genes of maternal origin motivating "other-directed" neocortical behaviors and genes of paternal origin motivating "self-directed" hypothalamic behaviors. First, uncertainty of paternity and predominant maternal care mean that social ties will often be stronger among maternal half-siblings than paternal half-siblings. Second, male-biased dispersal at reproductive maturity means that most mammalian social groups are based on matrilineal kinship (Haig 2000a), although humans may be a partial exception (Haig 2010b, 2011b).

For current purposes, I merely want you to entertain the possibility that maternal and paternal genes have conflicting interests, so that I can ask how such conflicts might be expressed within the mind. Imprinted genes would be expected to influence broad behavioral tendencies and personality traits, rather than micromanaging every individual decision. As Ishmael is tempted to betray his father, his genes cannot be well-informed about the details of his particular dilemma. Rather, his genes would have instinctual information about outcomes of similar choices in the past when more or less weight had been given to the recommendations of temptation versus conscience. If, for example, maternal genes had benefited, on average, when relatively more weight was given to temptation, then maternal genes would be expected to evolve adaptations to promote a greater tendency to succumb to temptation. But these adaptations of maternal genes would be opposed by adaptations of paternal genes to increase the suasive power of conscience. This internal tension could be played out during development, with maternal and paternal genes favoring growth of different brain structures, or it could be played out during brain function, with one set of genes enhancing, the other dampening the amplitude of particular neural signals.

What are the internal factors that determine our choices? Clearly, our natures are strongly influenced by the unique set of genes we inherit from our parents, but they are also influenced by the beliefs and memories we have accumulated during our lives. Our choices are influenced by ideas just as much as by genes. Most choices are simple, but we sometimes need to choose in situations when genes and ideas provide no clear guidance or when different internal voices offer different advice. Conflicts can exist among ideas, among genes, and between genes and ideas. The self can be viewed as the entity that is held responsible for our choices. We are free actors at least in the limited sense that no single set of interests exclusively determines our choices. We are also free in the sense that no one, not even ourselves, can predict with complete accuracy how we will choose in all situations.

7 Scratching Your Own Back

The previous chapter viewed the individual as a society of competing interests and posited an internal politics of the self. The study of human history reveals that polities may be characterized by periods of relative harmony, with effective social action to achieve common goods, interspersed with periods of bitter conflict, sometimes descending into civil war in which public goods are squandered. During periods of social peace, how are different factions able to work together despite divergent ideas about the just society? In the outbreak of civil strife, why does compromise collapse into internecine hostilities? When we come to consider the society of the self, the Hobbesian war of all against all is not easily resolved by submission to a Leviathan with a monopoly on force. We are a republic with dispersed centers of power rather than an absolute monarchy or dictatorship. How then do we achieve the benefits of cooperation and avoid the dangers of civil war? Can we come to internal compromises to which all parties accede? How do we resolve internal conflicts? Can we agree to disagree within ourselves? This chapter explores the possibility of strategic negotiation between contending parties in an internally complex self and addresses the question whether there can be intrapersonal reciprocity (Trivers 2011).

Introspection suggests that I often attempt to modify my own behavior by an internally voiced mixture of exhortation, bribes, and threats. Such intrapersonal cajolement has limited effectiveness, in part, because threats and promises to myself lack credibility. If I renege on a contract with myself, who will enforce sanctions? If I offer myself a reward now—for later good behavior—why am I bound to fulfill the second half of the bargain? Is it credible that I would knowingly do harm to myself to punish a past transgression? These attempts at intrapersonal persuasion pose a philosophical conundrum. I know what I know and I know what I want. So, why do I need to persuade myself? One might argue that internal bargaining is simply a reuse, or misapplication, of tools that are effective in the control of others' behavior. I modify my behavior in response to the threats and bribes of others and, in turn, use threats and bribes to modify others' behavior. Why not use these same methods on myself? Or, one might argue that intrapersonal bargaining resembles interpersonal bargaining because the self contains multiple agents with different, sometimes conflicting, agendas (Ainslie 2001).

Evolutionary conflicts arise within organisms when different replicators have different rules of transmission. Could these conflicts underlie problems of self-control? The potentially competing agendas within the self, both genetic and memetic, are complex. Therefore, to focus discussion it will be useful to consider a simple exemplar of an internal genetic conflict. For this purpose I will consider the conflict discussed in the previous chapter between the genes an individual (call him Bob, in honor of Robert Trivers) inherits from his mother and father.

Suppose that Bob's parents divorce and remarry, and that Bob's mother and father each have a child with their new partner. Bob thus has a maternal half-sibling (Maddy) and a paternal half-sibling (Paddy). What cost (C) would Bob's genes be prepared to accept in exchange for a benefit (B) to Maddy? Bob's maternal genes each have one chance in two of having being transmitted to Maddy. Therefore, his maternal genes would favor Maddy receiving the benefit, as long as the value to Maddy exceeded twice the cost to Bob ($B > 2C$). However, Bob's paternal genes are absent from Maddy because Bob and Maddy have different fathers. For Bob's paternal genes, no benefit to Maddy (no matter how great) could justify any cost to Bob (no matter how small). Bob's maternal and paternal genes are in conflict whenever $B > 2C > 0$. A similar conflict exists over benefits that Bob could confer on Paddy, but in this case the roles of maternal and paternal genes are reversed.

How could Bob's internal conflict be resolved? The traditional answer invokes a "veil of ignorance" (Rawls 1971). If genes lack information about their parental origin, they are constrained to behave in the same manner when inherited from either parent. If one of Bob's genes is faced with a choice whether to help Maddy, the gene is equally likely to be paternal, and necessarily absent from Maddy, or maternal, with one chance in two of being present in Maddy. Therefore, an uninformed gene has an expectation of one chance in four of its copies being present in Maddy and would therefore favor transfer of the benefit if $B > 4C$.

What theoretical predictions can be made about how the conflict would be resolved if the relevant genes are imprinted and thus possess information about their parental origin? Genes of both parental origins would favor withholding the benefit from Maddy if $B < 2C$ and both sets of genes would favor conferring the benefit if this directly benefited Bob ($C < 0$), but specific predictions of outcomes within the zone of conflict defined by $B > 2C > 0$ require assumptions about the mode of gene action and the relative power of the contending parties to influence the decision.

One possible resolution of a conflict is that one party has the power to dictate the outcome. If Bob's paternal genome had the dictatorial role, the benefit would be conferred only if it was without cost for Bob ($C < 0$), whereas if Bob's maternal genome had the dictatorial role, the benefit would be conferred when­ever the benefit to Maddy was more than twice the cost to Bob ($B > 2C$). In the first case, Bob's paternal genome could be considered to have a veto as to whether the benefit is conferred, whereas, in the second case, Bob's maternal genome could be considered to present his paternal genome with a fait accompli (Haig 1992b). Whether a particular dictatorial decision is interpreted as a veto or fait accompli may simply depend on how the question is framed. Models of gene expression at an imprinted locus result in a dictatorial outcome when maternal and paternal alleles have conflicting interests over the level of gene expression. At evolutionary equilibrium, whichever allele favors the higher amount produces that amount (the "loudest-voice-prevails" principle).

Stalemates could result if different genes had dictatorial power over different aspects of a decision-making process. Wilkins and Haig (2001) provide an example of this process in a two-locus model of the coevolution of a paternally expressed demand enhancer and a maternally expressed demand inhibitor. Increased paternal production of the demand enhancer favors increased maternal production of the demand inhibitor, which favors increased paternal production of the demand enhancer, and so on, until the evolutionary spiral comes to a halt because any further escalation would be prohibitively expensive for both parties. At evolutionary equilibrium in this model, the loudest voice prevails at both loci and there are substantial costs of conflict.

A possibility that has received little attention is that maternal and paternal genes could reach a compromise between their competing interests and avoid the full costs of evolutionary escalation. Suppose that separate opportunities arise for Bob to confer benefits on Maddy and Paddy. If the two opportunities were considered separately, the inclusive fitness of Bob's paternal genes would suffer if the benefit were conferred on Maddy, whereas Bob's paternal genes would suffer if the benefit were conferred on Paddy. Therefore, if Bob's maternal and paternal genes each had an effective veto on decisions whether to confer a costly benefit on half-siblings, neither benefit would be conferred. However, if the two decisions were considered jointly, both sets of genes might be able to increase their fitness if they could strike a deal. In effect, Bob's maternal genes could offer his paternal genes a bargain: "We will let you help Maddy if you let us help Paddy." The value of the deal for both parties would be $B/2 - 2C$. That is, each of Bob's genes experiences the costs of conferring the benefit on both Maddy and Paddy, in return for an expected half-share in the benefit to Maddy, but no share in the benefit to Paddy (or vice versa). If $B > 4C$, both sets of genes would receive a net inclusive fitness benefit relative to what they would achieve if they exercised their respective vetoes. If the deal can be trusted, it should be accepted.

In the above example, genes that know their parental origin settle on the same decision rule $(B > 4C)$ as genes that are uninformed about their parental origin. This is a consequence of artificial symmetries in the model and the limitation of bargaining to a single take-it-or-leave-it deal. Suppose instead that Maddy and Bob live together with their mother in Melbourne but Paddy lives with Bob's father in Dublin and never interacts with Bob. Bob's only opportunities to benefit a half-sibling involve Maddy. Bob's maternal genes have no opportunities to exercise their veto on benefits to Paddy and, therefore, have nothing to offer in exchange for Bob's paternal genes withholding their veto on benefits to Maddy. In this scenario, uninformed genes would use the decision rule $B > 4C$, but those of Bob's paternal genes that are informed of their parental origin have dictatorial power and should exercise their veto whenever $C > 0$.

Now consider an intermediate case in which there are two opportunities to benefit Maddy (combined benefit to Maddy $= 2B$; combined cost to Bob $= 2C$) for each opportunity to benefit Paddy (benefit to Paddy $= B$; cost to Bob $= C$). In one possible deal ("two-for-one"), Bob's paternal alleles would withhold their veto on both benefits to Maddy in exchange for Bob's maternal alleles withholding their veto on the single benefit to Paddy. The value of this deal to Bob's maternal alleles is $B - 3C$ whereas its value to Bob's paternal alleles is $B/2 - 3C$. Bob's maternal alleles would accept the deal if $B > 3C$ whereas his paternal alleles would accept the deal if $B > 6C$, but reject the deal if $B < 6C$. However, a deal in which one benefit to Maddy is exchanged for the benefit to Paddy ("one-for-one") is also possible. As shown above, this deal would benefit both parties (relative to no deal) if $B > 4C$. If $B < 4C$; neither deal is acceptable to Bob's paternal alleles who should exercise their veto. If $6C > B > 4C$, the "one-for-one" deal is acceptable to both parties, but the "two-for-one" deal is unacceptable to Bob's paternal alleles. "One-for-one" should be accepted by both parties. If $B > 6C$, either deal is superior to no deal for both parties, but Bob's paternal alleles would prefer "one-for-one" whereas his maternal alleles would prefer "two-for-one." Theory provides no simple prediction as to which deal, if any, will be chosen. The complications of multiple possible contracts are exacerbated if costs and benefits are continuous rather than discrete.

Genes of course are not subject to binding contracts. In the absence of enforceable contracts, the proposed deal—trading benefits to Maddy and Paddy—has the nature of a prisoner's dilemma. If the effects of genes are fixed and not conditional on the behavior of the other genes with which they are temporarily associated, the deal is a one-shot prisoner's dilemma and is evolutionarily unenforceable. Although mutual defection is the evolutionary stable strategy (ESS) if a prisoner's dilemma is played once with a particular partner ("I stab your back, you stab mine"), cooperation can sometimes succeed if the game is played repeatedly with the same partner and there is uncertainty about when the relationship will end (Axelrod and Hamilton 1981). If this criterion—long-term association of uncertain duration— were the only consideration, strategic cooperation among genes within an organism should be possible.

Most successful cooperative strategies in iterated prisoner's dilemmas (IPDs) require memory of past interactions to allow a player's actions to be conditional on the past behavior of its partner. If genes were to implement a simple strategy like tit for tat ("Do what the other player did on the previous move"), they would need to perceive, and remember, a partner's previous move and use this information to guide current behavior. This requires a degree of sophistication that some may be wary of ascribing to genes, even though genes are known to possess forms of memory and to exhibit complex conditional expression. One might ask, however, why cooperation should be more difficult to evolve for interactions between genes within organisms than for interactions between genes in different organisms.

Analytical investigations and computer simulations of the IPD have revealed a wealth of possibilities for complex behavior. In general, no strategy is an ESS and multiple strategies may coexist in a population. Polymorphism is possible because two strategies may be indistinguishable when played against each other, so that their relative performance will be determined by their interactions with third strategies, including the degenerate versions of themselves produced by mutation (Bendor and Swistak 1997). In the context of intrapersonal reciprocity, genetic polymorphism would raise the intriguing possibility that different combinations of strategies in the internal IPD could contribute to personality differences: sometimes one interest gets its way; sometimes the different interests come to a relatively harmonious compromise; sometimes the different interests contest every point of contention.

It is tempting to simply transfer results obtained from theoretical investigations of the IPD to discussions of reciprocity within the genome. I resist the temptation because most simulations of the IPD have competed strategies in a stereotypical tournament (Bendor and Swistak 1997). In each round of the tournament, pairs of strategies play multiple rounds of the prisoner's dilemma, with "fitness" determined by the sum of payoffs. In the next round, a strategy increases or decreases in frequency determined by its performance relative to the average fitness of the population. Significantly, a strategy is inherited as a whole and does not undergo recombination. This kind of tournament models asexual or single-locus inheritance of strategies.

Consider, however, an alternative tournament in which *teams* play against each other, and in which the members of a team have different tasks in the collective implementation of their team's strategy. In this tournament, a team's relative performance in one round determines the probability that its members will participate in the next round, but the teams themselves are ephemeral. At the end of each round, members of successful teams (or their clonal progeny) are reorganized into new teams which reenter the tournament to play against another reorganized team. This kind of tournament models sexual or multi-locus inheritance of strategies, with the units of inheritance determining components of strategies rather than entire coherent strategies. A successful component must do well, on average, across all the combinations in which it is tested. The sexual tournament would appear to select for a team of champions rather than a champion team. It is possible that the simpler asexual tournaments provide a good guide to equilibrium behavior in the more complex sexual tournaments, but this is something to be demonstrated rather than simply asserted.

Speculation about intrapersonal reciprocity is fun, but prompts the question how one would ever know that it occurs. How does one distinguish strategic cooperation from veil-of-ignorance cooperation or from escalated noncooperative stalemates? The problem of testability has bedeviled progress in the scientific study of internal conflicts, because there may be little that we can observe in external behavior that allows us to test hypotheses about internal conflicts. Progress in this field is likely to come from understanding the details of genetic mechanisms, just as an appreciation of internal conflicts between paternal and maternal genes has in large part depended on increased knowledge about the molecular functions of imprinted genes (Wilkins and Haig 2003). One would need to demonstrate conditionality of gene expression, not just expression that is conditional on parental origin, but also expression that is conditional on the behavior of other genes. This is a challenge, but surely not an insurmountable one.

Afterthoughts

> For the flesh lusteth against the Spirit, and the Spirit against the flesh: and these are contrary the one to the other: so that ye cannot do the things that ye would.
>
> —Galatians 5:17, King James Version

We often have the subjective experience of struggling with ourselves, of a conflict between powerful internal voices in which neither side yields nor gives up the fight, in which neither side plays fair. Elsewhere I have argued that the multiple "factions" within the genome tend to align into two opposed "parties" and that the hardliners of each party often drive the debate on substantive issues (Haig 2006). Many diagnoses of the etiology of our internal conflicts invoke two opposing forces, variously described as a struggle between impulse and control, passion and reason, temptation and conscience, sin and virtue. There is a family resemblance in the axes of dispute in these diagnoses, perhaps analogous to the first principal component in a multidimensional space of conflicting forces. And the conflict of these forces is sometimes governed by an executive controller or judge. Examples include Socrates's simile of a tripartite soul, a charioteer struggling with a dark and white horse, or Freud's *Ich* mediating the conflict between *Es* and *Über-Ich* (egregiously translated as *ego*, *id*, and *superego*).

Why do some minds function harmoniously whereas other are consumed in recrimination and internecine strife? The standard biological approach to mental illness is to search for a broken mechanism; but could some discontents of the self be conceptualized as an expression of a dysfunctional community rather than a malfunctioning machine? Could such a shift in perspective help to heal troubled souls? I have been timid about speculating on the implications of internal conflicts for human psychopathology because it would be easy to be misunderstood as advocating crude genetic determinism and to be accused of being insensitive to nuance and of failing to understand the "insuperable" divide between mechanisms and subjective experience.

One issue raised in this chapter is whether we can credibly threaten ourselves to coerce our own behavior. Could a "faction" within the self follow through on a threat in order to make future threats credible? "Self-destructive" behaviors immediately come to mind. Could these behaviors sometimes result from a breakdown of reciprocity among the factions of self, a fraying of an internal social contract, with the infliction of actual harm in a desperate attempt to change future behavior? For those of us who have known individuals who have engaged in self-mutilation of body or mind, the behaviors can be hard to fathom. From my perspective as a completely unqualified observer, the outward signs seem to express a degree of anger of a faction within the collective self toward the self, combined with a desire to prove to the self the strength of the faction's resolve. But this may be a failure of empathy on my part, with an exaggerated projection onto others of my own inward anger when the "meddling" of one internal faction "frustrates" the plans of another.

8   Reflexions on Self

This chapter descends from an essay written for a joint celebration of the 250th anniversary of *The Theory of Moral Sentiments* and the 150th anniversary of *On the Origin of Species by Means of Natural Selection* and contains the reflections of one twenty-first-century Darwinist on the passions, reason, and morality, inspired by his rereading Adam Smith's eighteenth-century masterpiece. Smith centered his account of the moral sentiments on the concept of sympathy: we understand others by putting ourselves in their situation; and we judge our own conduct by viewing ourselves from the perspective of an impartial spectator. Our faculties of sympathy are both reflexive, autonomic responses beyond the control of our wills, and reflective, reasoned contemplation of others and our relations to them. Smith's prose matches his subject, adopting multiple perspectives and switching voices, at times intimate and passionate, and at others distant and reserved. There is a rhythm to his prose that resonates in the mind of the reader, as well as, to my mind, a playful seriousness that invites a serious playfulness in reply. *The Theory of Moral Sentiments* is not only a work on sympathy but also a work that evokes a sympathetic response in the reader.

My essay attempted an explication of the moral sentiments that melded insights of Darwin and Smith. The present chapter has inherited the form of three main sections. The first discusses the multiplicity of guides to individual action under the broad categories of instinct, reason, and culture, and discusses different kinds of answers to the question why we act the way we do. The second explores different kinds of reflections (and reflexions) back upon our internal self-image. The third uses the arguments of the preceding sections to discuss the nature of our moral faculties. Our moral choices are viewed as emerging from a nexus of conflicting agendas of different entities with different ends. I will suggest that a locus of moral responsibility, and a sense of self itself, emerges as we flip back and forth between our own perspective and the perspectives of others, and as we attempt to reconcile and adjudicate among the different springs of internal action.

In thinking about sympathy, my mind kept returning to the metaphor of mirrors reflecting mirrors: we see ourselves through others' eyes, who see themselves through our eyes. In keeping with this theme of reflection, and reflections upon reflections, I resolved to give my essay a recursive structure in which the text constantly reflected back upon itself. As a work of sympathy with Adam Smith, I did not attempt to achieve complete clarity in the text, nor within myself, about when I spoke in Smith's voice and when I spoke for myself. Such ambiguity seems fitting when discussing a topic that blurs the boundaries between individuals and their not-so-distinct points of view. My views depart most from Smith's treatment of sympathy when I allude to the usefulness of sympathy in manipulating and exploiting others for selfish ends. Perhaps Smith did not consider that instrumental uses of sympathy came under the purview of the *moral* sentiments; or perhaps he had more faith than I in the beneficence of creation.

Guides to Action

> In every part of the universe we observe means adjusted with the nicest artifice to the ends which they are intended to produce; and in the mechanism of a plant, or animal body, admire how every thing is contrived for advancing the two great purposes of nature, the support of the individual, and the propagation of the species. . . . But though in accounting for the operations of bodies, we never fail to distinguish in this manner the efficient from the final cause, in accounting for those of the mind we are very apt to confound these two different things with one another. When by natural principles we are led to advance those ends, which a refined and enlightened reason would recommend to us, we are very apt to impute to that reason, as to their efficient cause, the sentiments and actions by which we advance those ends, and to imagine that to be the wisdom of man, which is in reality the wisdom of God.
>
> —Adam Smith (1976)

Smith recognizes two levels of causal explanation in this passage: our reasons and the reasons for these reasons. The specific question to which he alludes is the source of our approbation of the punishment of individuals who violate moral laws. In Smith's opinion, we punish, and approve of punishment, out of our indignation against the offender, not because of a reasoned consideration of the value of punishment for the proper ordering of society. Our indignation, however, has been contrived as an efficient means of advancing the latter end. We act out of passion, but the preservation of society is the reason why we have been endowed with this passion.

Smith's grounding of final causes in God's wisdom could be considered an orthodox appeal to natural theology, but his ontological stance on the nature of final causes is not altogether clear. When he wants to, Smith writes clearly, and his lack of clarity on teleological questions is, I suspect, deliberate. A hundred years later, Darwin provided a naturalistic account of the appearance of purpose in nature: spontaneously arising variation modifies the properties of organisms; some of these modifications benefit the organism in its struggle for existence, and these modifications are thereby perpetuated in the organism's offspring; thus, an *effect* of a modification of the organism is a *cause* of that modification appearing in subsequent generations.

Darwin's understanding of the hereditary material was inchoate. He would have accepted support of the individual, and even the tribe, as the "goal" of the adaptive process, but many now prefer to view the genetic material itself, rather than the individual or group, as the beneficiary of the fruits of natural selection (Dawkins 1982). This is a subject of ongoing debate in the philosophy of biology, with most of the polemical heat concerning semantic rather than substantive issues. From a gene-centric perspective, a gene's functions (or purpose) are those of its *phenotypic effects* that have a *causal role* in the gene being preserved and propagated.

The recognition of different levels of explanation is familiar to evolutionary biologists. Mayr (1961) distinguished proximate *How?* explanations from ultimate *Why?* explanations. Tinbergen (1963) recognized four kinds of explanation: physical causation, survival value, evolutionary history, and ontogeny. I prefer to treat psychological *motivation* as a fifth kind of explanation, complementary to the others, rather than as a special kind of proximate mechanism or physical cause. When I wish to understand *why* you behaved as you did, I am usually asking a question about the *telos* of your motivations. When I wish to persuade you to do something you might not otherwise do, I am interested in *how* your motivations can be used as means to the ends of my motivations.

Consider a serial philanderer who copulates with multiple women by making false promises of commitment. He does not copulate to pass on his genes, but to experience sexual pleasure at little personal cost. The sexual gratification he receives after each successful seduction serves to reinforce the seductive behaviors (intrapersonal recursion). But, the system of sexual desire, seduction, gratification, and reinforcement exists, in part, because he had ancestors who passed on their genes because they consummated their desires by seduction (evolutionary recursion). The philanderer copulates to experience sexual pleasure, not to transmit his genes; but copulation is pleasurable because the promise of pleasure has been the means whereby our ancestors were induced to copulate.

Two levels of *teleology* can be recognized in this example. Sexual desire and the behaviors it motivates are the means that achieve the end of sexual pleasure for the philanderer, and the means that achieve the end of open-ended replication for his genes. The observation that the philanderer prefers that copulation not result in conception, whereas his genes "prefer" that it does, clearly illustrates that psychological motivation and evolutionary function are distinct. The philanderer may use a condom and thwart his genes' ends. Or, his genes may promote "irresponsible" behavior that conceives an *unwanted* pregnancy.

A failure to distinguish evolutionary function from psychological motivation bedevils many debates about our moral nature. Are we ultimately selfish, or do we genuinely care for others? Both could be right. Benevolent motivations would not cease to be benevolent if they were shown to be adaptations of selfish genes.

Instinct

> With regard to all those ends which, upon account of their peculiar importance, may be regarded, if such an expression is allowable, as the favourite ends of nature, she has constantly in this manner not only endowed mankind with an appetite for the end which she proposes, but likewise with an appetite for the means by which alone this end can be brought about, for their own sakes, and independent of their tendency to produce it.
>
> —Adam Smith (1976)

Fitness is the *telos* of our genetic adaptations, but each passion also has a proximate *telos* toward which it cajoles us to action. Hunger has the goal of food; thirst, the goal of water; lust, the goal of sexual gratification; and the desire for breath, the goal of oxygen in the life-saving gasp of a drowning man. Our social desires have social goals: to be accepted, to be admired, to be loved, to be feared, and to be revenged. For want of a better term, I will refer to the proximate *telos* of each passion as its *utility*. More utility is preferable to less utility for each passion, but I do not wish to presuppose that the utilities of different passions are commensurable or representable on a single scale.

Utilities are correlated with fitness, but fitness and utility are not the same thing. Children, not genes, are the proximate goal of the desire to have children, a *telos* that is sometimes attained by adoption or ovum donation. Passions may misfire and fail to enhance fitness, but, from our personal perspective, happiness is happiness whether or not our genes benefit. Genes have no "preference" whether fitness is achieved through our misery or our happiness, through our hatred or our love. But we may have distinct preferences among the alternative paths urged by our passions.

Our instincts can be used as means to others' ends. This is particularly clear when an advertiser uses our hopes and our fears to enrich himself and anonymous shareholders, but the same is true of all forms of persuasion. When the Catholic church persuades a young man to enter the priesthood and forsake genetic reproduction for a religious calling, his motivations are used to serve the ends of a cultural tradition. Nevertheless, the priest may attain the proximate ends of at least some of his passions if he leads a happy and contented life; and he may serve the ends of his moral code as well as the ends of believers' genes if he persuades his parishioners to have larger families.

Genes "learn" the hard way, by the selective elimination of the less fit. But one of the most useful lessons "learned" by genes is how to construct organisms that learn from experience and thus compensate for the limitations of innate mechanisms. We learn from the mismatch between our projected and actual performance; from the repetitive performance we call practice; from the mock performance we call play; and from the virtual performance we call reasoning to consequences. And we learn from the performance of others. Our faculty of imitation allows the costs of learning to be shared among individuals and expertise to accumulate over generations. By these means, we come to embody more adaptive information than is present in our genome. We are rational and cultural beings, instinctively. But reason and culture can employ our passions in pursuit of ends other than fitness.

Reason

Reason encompasses innate problem-solving mechanisms that may be conscious or unconscious and can be employed in the pursuit of multiple ends. Reason can respond to novel features of our current environment, for which there has been no past natural selection, and can take advantage of small differences in expected return that constitute too weak a selective force for the evolution of hardwired adaptations.

Reason is both the slave and the governor of the passions. As a slave, reason is used to find effective ways to further the objectives of the passions. In the process, reason may identify subsidiary goals on the path to satisfying a desire. If I am thirsty, I need to find a safe route into a ravine or a quick way to improvise a bucket. In this manner, reason modifies and gives specificity to the goals of our passions. We have multiple passions because fitness is achieved in different ways in different contexts. Actions that enhance fitness in one context sometimes will have been disastrous in another context. Therefore, reason, as a governor, has evolved to reconcile and adjudicate among the passions in the context of a specific life, in a specific environment, and a specific culture.

The *telos* of reason—the "utility function" that compares and aggregates the preferences of different passions—remains for me something nebulous and ill-defined. The theory of social choice may provide a better model than the theory of rational choice of what goes on within the mind if reason cannot make intrapersonal comparisons of utility when aggregating preferences of different agents within the self. And this theory suggests that basic precepts of normative rationality may be violated if cardinal utilities cannot be compared. Introspection fails to reveal clear criteria by which I resolve internal dilemmas. My decisions are unstable and seem to aim at a *reasonably* happy life while juggling duties and obligations to other individuals, all in the context of conflicting cultural suggestions on how best to balance the passions to achieve a "good life" (whatever that may be).

I will leave such foundational questions to one side, assume that reason recommends an action opposed to the blandishments of some passion, and ask by what principles such an internal conflict could be resolved. The conflict between passion and reason is often posed as a problem of self-command or strength of will. From a genetic perspective, this is a conflict between two adaptations. Passion summarizes the wisdom of past natural selection and recommends actions that have promoted fitness in similar situations in the past, whereas reason responds to specific features of the current context and can potentially recognize whether emotion recommends actions that are inappropriate and self-defeating in this particular instance. But reason too is fallible and its choices can be influenced by the persuasion of other actors who may not have our genetic interests at heart. Neither reason nor passion can be guaranteed to give better advice in any particular case. Strength of will is likely subject to stabilizing selection: too weak a will, and passion will have its way on many occasions when reason is wiser; too strong a will, and reason will too often ignore the conservative wisdom of passion.

Reason can be a passion of its own. Solving puzzles is rewarding in itself. The desire to make sense of the world is a psychological motivation that is consummated in the pleasure of understanding and the thrill of discovery. We never know what knowledge may be useful to achieve future goals, and the solutions we discover to life's conundrums are "intellectual property" that we can share or trade with others for our advantage.

Culture

We learn from the experience and reasoning of others and thereby reduce the costs of individual trial-and-error learning. We exercise discrimination in what we learn and tend to emulate the behavior and opinions of those who have achieved, or seem to have achieved, the things we desire in life. Our teachers themselves had grandteachers who themselves had great-grandteachers, with discrimination and recombination exercised at each link in this chain of transmission. By this cumulative process, the ideas that are culturally transmitted evolve to become ever more attractive to learners. Cultural traditions acquire adaptations that promote their own transmission and thereby possess their own teleology. Culture has a tendency to enhance the utility and fitness of its carriers because we adopt cultural items that appeal to psychological motivations that have been shaped by natural selection. But culture evolves much more rapidly than the passions and by processes that need not promote either fitness or utility (Richerson and Boyd 2005). And thus, our psychological motivations become partially untethered from the moorings of their genetic function.

Sympathy and the Proliferation of Selves

> As we have no immediate experience of what other men feel, we can form no idea of the manner in which they are affected, but by conceiving what we ourselves should feel in the like situation.
>
> —Adam Smith (1976)

Sympathy—the vicarious experience of others' actions, emotions, tastes, and reasons—is central to Adam Smith's account of the moral sentiments. In writing this chapter, I entered into sympathy with his thought and his rhetoric, and saw traces of his cadence and his style creeping into my thought and my prose. By identifying with him, I developed affection for the long-dead philosopher, and wish him well, but benevolence is not a necessary part of my definition of sympathy. We also sympathize with others the better to use them as means for our ends, and to avoid being exploited as means for their ends. Notice how my prose slipped from the first-person singular, in talking of *my* benevolence, to the first-person plural, in discussing *our* less attractive motives. By so doing, I attempted to engage your sympathy to temper your judgment of my sins.

Mirror neurons are commonly interpreted as the neurological basis for our sense of sympathy (Fogassi 2010; Gallese 2007; Molnar-Szakacs 2010), but I am no neurologist, and, in emulation of Smith, I will write in generalities and abstractions. Below, I recognize three levels of sympathy. First-person sympathy involves identification with our own self-image and is the scaffold upon which we construct images of other selves. Second-person sympathy involves experiencing the viewpoint of a person with whom we are directly interacting. It facilitates cooperation via direct reciprocity: *I behave well by you so that you will behave well by me.* Third-person sympathy involves viewing our own conduct from the perspective of an impartial observer. It facilitates cooperation via indirect reciprocity: *I behave well by you so that others will behave well by me.*

First-Person Sympathy

We possess a first-person mental image of the current position of our members that is crucial in planning motor actions. Integral to this self-image is perceptual feedback that allows future plans to be modified by past performance. When we miss a target with a thrown projectile, we adjust our posture and point of release at our next attempt. The overlapping representation of action and perception of action allows us to learn from experience and to perfect our aim, by bringing the predicted perception of action into register with the actual perception of action.

Our mental self-image keeps track of many aspects of our physical and emotional state besides the position of our limbs, helping us to adjust, understand, and predict our own behavior. We enter into sympathy with ourselves (first-level recursion) and "share" in our own hopes and fears. Just as each of us can improve, if not perfect, our jump-shot through self-directed effort, we each have a limited ability to set ourselves moral goals and become more like the person we wish to be.

Second-Person Sympathy and Direct Reciprocity

> When we see a stroke aimed and just ready to fall upon the leg or arm of another person, we naturally shrink and draw back our own leg or our own arm. . . . The mob, when they are gazing at a dancer on the slack rope, naturally writhe and twist and balance their own bodies, as they see him do, and as they feel that they themselves must do if in his situation.
>
> —Adam Smith (1976)

We use our image of our self as the template on which we construct images of other selves, and are thereby better able to understand and predict their actions and to learn from their experience. When we watch a television chef whisk an egg, we utilize her experience, and the experience of her teachers, to improve our soufflé, in part by direct bodily simulation of her movements. Bodily sympathy is thus one of the essential components of our faculty of culture. Not only does our second-person image of another allow us to learn from her experience but it also allows us to interpret her feelings, and predict her actions, for our benefit, whether we use this knowledge for her good or her ill. Second-person sympathy can be reflexive or reflective. When I watch a dancer on a slack rope my body *naturally* writhes and twists as I feel I *must* do in his situation, but when I enter into the mind of an opponent at chess, I attempt to match my reason and self-command to his.

My second-person image of you incorporates my perception of your current state shaped by my memories of your past behavior. This representation helps me to predict your future behavior in our current relations. Do I *feel* you are somebody who can be trusted, or somebody it would be foolish to trust? Do I *feel* you are somebody who can be exploited without fear of reprisal, or somebody it would be dangerous to cross?

Second-person sympathy is simulated experience, not actual experience, and images constructed by the eye of sympathy may be distorted, or out of focus, in a manner analogous to imperfections of the images constructed by our outer eyes. Natural selection favors increased acuity of both inner and outer vision, but this does not preclude occasional sympathetic myopia or cataracts of compassion. My first-person model of myself may be a poor prototype for constructing my second-person model of you because our characters differ, because of failures of my imagination, or because my self-image is itself distorted.

Although our first-person image of ourself is the template for our second-person images of other selves, we are not restricted to viewing others as identical to ourself. We can learn their idiosyncrasies and model their behavior as somewhat different from our own: a skilled boxer anticipates and deflects the punches of

both left-handed and right-handed opponents; a skilled philanderer embodies knowledge of female nature, and of individual
women, the better to compromise their virtue.

It is relatively simple to adjust for the lesser abilities of others in our sympathetic appreciation of their situation, but difficult, if not impossible, to simulate abilities we do not possess
ourselves. A chess master can lure an average player into a trap
by viewing the board through his opponent's eyes, but the average player flounders in his attempt to anticipate and block the
moves of the master. Natural selection thus favors an escalation
in our abilities, including our sympathetic abilities, the better
to manipulate others and to avoid being manipulated by them.
But we are less likely to trust those with whom we cannot sympathize and thereby find less predictable. Natural selection may
also dampen the development of unusual abilities because of our
need to live and be accepted in society.

Reflected First-Person Sympathy

Second-person sympathy creates recursive complexity. My
second-person image of you includes my simulation of your
second-person image of me, and I project this *reflected image*
onto my first-person image of myself (second-level recursion).
If I feel you feel warmly to me, I feel more warmly to myself. If
I feel you find me unattractive, I feel less attractive. I become
more aware of myself by becoming more aware of you, and my
first-person sympathy is thereby enriched. But if I have a distorted image of myself, my image of your image of me reflects
the distorted image back upon itself, further distorting the original image. If I project a deceptive image of myself to others, I
risk becoming a victim of the same deception via the reflected
image of myself. Feedback from reflected self-images must be

appropriately damped if our self-image is not to develop extreme distortions or undergo wild fluctuations. One might refer to such aberrations as *sympathology*.

The reflected image of myself provides an answer to the questions: Do I *feel* you feel I am somebody who can be trusted? Do I *feel* you feel I am somebody who can be exploited? In principle, my second-person image of you also includes my simulation of your simulation of my second-person image of you (third-level recursion). This image of you, in the reflected image of me, allows me to answer the question: Do I *feel* you feel I feel you are somebody who can be trusted? My difficulty in mentally parsing higher terms of the series indicates cognitive limits on my depth of recursion that may, in turn, reflect the limited adaptive value of deep introspective self-awareness. Perhaps you find it easy. But then, this is something about you I cannot understand, beyond that you tell me it is so.

The fact that you form a second-person image of me that influences how we interact provides me with an incentive to manage that image. As a result, my first-person representation of myself needs to keep track of the things that I present to your perception—a *projected self* that I intend should mold your second-person representation of me—and the things I hide from your perception—a *private self* that I do not wish to influence that image. As a further consequence, my second-person representation of you will include an estimate of your private self, constructed from things you have unintentionally given away and from guesses about what you hide from me, based on inferences from what I hide from you. Perhaps there are feelings or thoughts I hide from myself, that are excluded even from the image of my private self, the better to hide them from you (Trivers 2000, 2011). And, perhaps, in some cases, we both benefit from some motivations not forming a part of our reflected self-images.

We construct different projected and private selves for different others, one for our mother, one for our spouse, and potentially one for each individual with whom we interact. But in practice, we develop a catch-all projected self, a *public self*, that we present to most others. This consistent self minimizes both the cognitive load of keeping track of a proliferation of self-presentations and the risk of being detected in duplicity by third-party observers. The larger the mismatch between our private and public self, the greater the risk of detection, and the less useful the feedback we receive from our reflected self-images. Integrity is born of prudence.

Third-Person Sympathy and Indirect Reciprocity

> Before we can make any proper comparison of those opposite interests, we must change our position. We must view them, neither from our own place nor yet from his, neither with our own eyes nor yet with his, but from the place and with the eyes of a third person, who has no particular connexion with either, and who judges with impartiality between us.
>
> —Adam Smith (1976)

A particularly important contribution to our second-person images of others was alluded to at the end of the previous section. You and I may bring to our current interaction second-person representations of each other that have been shaped by observation or report of each other's past interactions with third parties. Moreover, a third-party observer of our current interaction may form a second-person representation of me that will shape his behavior in future interactions with me, for my good or my ill, and which he may share with parties unknown. Therefore, in my interaction with you, I need to maintain and monitor a second-person representation of an observer that includes his second-person representation of me. I must see myself through his eyes and sympathize with his approbation or disapprobation of my character. I must mold my public self to his judgment. Unlike previous second-person images, this representation of someone who might be watching is not a model of a specific individual, but of a generic observer. Its most salient feature, prominent in an otherwise nondescript image, is the reflected image of ourselves as we might appear to others.

We construct different third-person images of different kinds of observers. An elderly dark-skinned woman crying at the inauguration of Barack Obama felt a hint of her own acceptance in the president's acceptance by the nation, she re-experienced unjust judgments of the past, and perhaps adjusted the reflected image of herself as she adjusted her own representation of a light-skinned observer. Or so judged this light-skinned observer in my second-person representation of her first-person response to her third-person representation of an observer like me.

Man was made to live in society. He procures great benefit from interaction with his fellows and would sustain great cost if he were to lose their esteem and be excluded from their company. Although he may gain short-term benefits from exploiting others, he can rarely be sure no one is watching. The potential future costs from damage to his reputation, and his reduced ability to enjoy the benefits of society, will often dwarf any small advantage he might gain from misbehavior in the present. Thus, in man's mental model of the judgment of a third-party observer, we have arrived at a close relative of Adam Smith's "man within the breast," the impartial spectator who judges our conduct. Direct reciprocity—I behave fairly by you, so that you behave fairly by me—now shares the stage with indirect reciprocity—I behave fairly by you, so that others will behave fairly by me (Alexander 1987; Nowak and Sigmund 1998).

Smith wrote: "Nature, when she formed man for society, endowed him with an original desire to please, and an original aversion to offend his brethren" (1976, 116). In this chapter, I use "man" and masculine pronouns to refer to both sexes. This was the convention of Smith's day. It showed a lack of sympathy with female readers. I chose the generic masculine knowingly, to accentuate my sympathy with Smith, but with anxiety about how it would affect my second-person representation in your minds. On a final note, it is worth noting that Smith did not see third-person sympathy as incorruptible: "When he is at hand, when he is present, the violence and injustice of our own selfish passions are sometimes sufficient to induce the man within the breast to make a report very different from what the real circumstances of the case are capable of authorising" (Smith 1976, 159).

On Tensions between Second- and Third-Person Sympathy

Second-person sympathy involves engagement with the perspective of a particular other whereas third-person sympathy involves engagement with the perspective of a generic other. And these two others may view differently how I should act in a particular situation and present me with different reflected images of those actions. Virginia Held (2006) argues that moral philosophy has overemphasized the claims of third-person detachment and defends an "ethics of care" based on second-person engagement with particular others. Vasudevi Reddy (2008) seeks to redress a similar imbalance between third-person inference and second-person engagement in theories of how infants come to understand other minds.

A friend is someone with whom we develop intense second-person sympathy and with whom we partially disengage the third-person perspective. When a friend involves me, without asking, in a conspiracy against the common good, how should I reconcile my moral obligations to friendship with my moral obligations to society? At what point does my desire to remain in good standing with my friend give way to my fear of losing standing in society? And what does my dilemma tell me about my friend? Has he implicated me because he has a faulty second-person image of me (a lack of sympathy for the man within my breast), or because he has an accurate image and knows I will bend? My friend possibly views the opportunity for mutual gain that he offers me to be an expression of the value he places on our friendship and a means of strengthening and deepening that relationship. From his perspective, my moral qualms could be a sign that I care more about abstract principles, and the judgment of others, than I care about him. A person who is always looking over his shoulder, and considering how he appears to the world, can seem cold and detached. In our most intimate relations, we want a partner who is attentive to us, not the rest of humanity. In a pinch, I want someone who loves *me* and values my interests above others'.

Reflections on Rhetoric and Belles Lettres

The sole reader of the early drafts of this essay was the writer. His first attempts did not convince and were sent back for revision. In the process of reading his words, and observing his struggles to persuade, my ideas took form and, in a resonance between reader and writer, I came to know my own mind. I, the author, had a second person in mind. It was you, dear reader. If I have succeeded in reading through your eyes, then perhaps you will see through mine. You will have entered into sympathy with my thought, just as I have entered into sympathy with Smith's. Modern academic writers must also consider a third person: a reader who observes the second-person transaction between author and reader from one side, who does not engage directly with the author's intended meaning but with the tools and tricks the text uses to achieve its effects. What creates passionate engagement of the second-person reader of the author may not satisfy the reasoned judgment of the third-person reader of the text (Brown 1994).

Morality

> Though it may be true, therefore, that every individual, in his own breast, naturally prefers himself to all mankind, yet he dares not look mankind in the face, and avow that he acts according to this principle.
> —Adam Smith (1976)

Morality is not one thing but rather an imperfect amalgam of instinctive, rational, and cultural elements. As with all things evolutionary, the relations among these elements are recursive. Culture is shaped by instinct, and instinct by culture. Reasoned choices facilitate the evolution of innate behaviors that achieve the same end without reason (Baldwin 1896). Reason selects cultural items that best serve our ends, and culture is thereby changed, but reason's choices are constrained by cultural norms. Morality is both deeply personal and highly social, something that emerges from within and is imposed from without.

Instinctive Elements

Many aspects of morality are instinctive, modified by personal experience, reason, and culture. Our basic repertoire of emotional responses is innate: including the resentment and indignation we feel toward the selfishness of others, the gratitude we feel toward their generosity, and the guilt or shame we feel for our own trespasses. But our emotional repertoire also includes (among other unamiable passions) the envy, jealousy, and even hatred we feel toward those who possess something we want, and the desire for revenge and retribution when others thwart our goals. Our faculty of sympathy, our ability to see the world from another's perspective, is an essential substrate of morality, but sympathy coexists with *Schadenfreude*, pleasure that we feel in the pain of another, especially one who has wronged us.

Our moral instincts evolved because they promoted the differential survival and reproduction of our ancestors, relative to other individuals, with other instincts, who survived less well and produced fewer offspring. For most of our ancestors, leaving genes to posterity depended on being accepted into social groups that maintained collective control over their own membership. Individuals who engaged in "unacceptable" behaviors could be shunned, excluded from the enjoyment of public goods, expelled from the tribe, or even killed. Thus, the desire for acceptance and the dread of rejection are among the most powerful of human motivations.

Third-person sympathy is likely to have evolved because of the great benefits to be derived from within-group cooperation and the great cost of exclusion from those benefits. If reproduction were contingent on acceptance into a group, and acceptance was contingent on the expression of "moral" behaviors and the suppression of "immoral" behaviors, then humans may have socialized themselves by excluding carriers of genes for antisocial behaviors from their company.

If the prudent course is to seem to possess other-regarding preferences, then the most effective way to seem to possess such preferences may be to actually possess them. Prudence may have initially recommended the appearance of benevolence, but natural selection may then have favored genetic changes that conferred benevolence as a psychological motivation. The prudential reasons for benevolence could, in principle, be transferred entirely from the realm of utility and reason to the realm of fitness and instinct, leaving benevolence as a pure psychological motivation, untainted by self-interest. We could thereby look our fellows in the face and truly avow that we value their happiness as our own. In truth, self-interest remains a powerful psychological motivation, with self-love and benevolence coexisting in our psyches as instinctual motivations that are not fully reconciled.

The evolution of cooperation is often presented as the outcome of a tension between selection among groups and selection within groups (Sober and Wilson 1998). In this view, competition among groups favors cooperation within groups, with the latter constantly undermined by a reproductive advantage for freeloaders who share the benefits of cooperation but do not pay the costs. The two levels of selection work in concert to enhance cooperation, however, if groups expel their "antisocial" members or withhold the fruits of cooperation from those who do not contribute. Thus, the exigencies of between-group conflict may have been a powerful force for the enforcement of solidarity and moral policing within groups, if groups of cooperative individuals were better able to expand and defend their territories, sometimes by the brutal extermination of their neighbors (Wrangham 1999).

Natural selection may also have favored the psychological ability to negotiate truces and avoid mutual destruction in escalated conflicts, but only if the inability to bury the hatchet was associated with reduced genetic reproduction at some level of between-group or within-group competition. Instinctive peacemakers may inherit the earth, but only if they leave more descendants than instinctive grudge-holders. Both parties benefit from a truce, but each party must be wary of shifts in the balance of power that favor opportunistic aggression by the other, and, a corollary less often acknowledged, must be ready to exploit opportunities for low-risk aggression when they arise.

Rational Elements

But though reason is undoubtedly the source of the general rules of morality, and of all the moral judgments which we form by means of them; it is altogether absurd and unintelligible to suppose that the first perceptions of right and wrong can be derived from reason, even in those particular cases upon the existence of which the general rules are formed. . . . Nothing can be agreeable or disagreeable for its own sake, which is not rendered such by immediate sense and feeling.

—Adam Smith (1976)

Those general rules of conduct, when they have been fixed in our mind by habitual reflection, are of great use in correcting the misrepresentations of self-love concerning what is fit and proper to be done in our particular situation.

—Adam Smith (1976)

Smith believed our moral intuitions come from our passions, but our moral rules come from reason. We derive general rules not from the judgments of the man within our breast, a corruptible spectator who cannot always view our actions dispassionately and impartially, but from the natural sentiments we feel when we observe the actions of others and observe others' judgments of these actions. In particular, we desire the admiration that comes from general approbation of our conduct and we fear the opprobrium from general disapprobation. We derive our moral rules from the sentiments we experience in third-party observations of others, not from our simulated, and less-reliable, third-party observations of ourselves (Smith 1976, 156–160).

Reason may be used to *rationalize* our behavior and our opinions; to justify, to ourselves and to others, positions we have already adopted. Jonathan Haidt (2001) has argued that the principal use of reason in moral judgment is finding *post hoc* justifications for moral intuitions. In Haidt's view, moral reasoning has a limited power to change the behavior of others by reframing questions and triggering new moral intuitions, but it is usually ineffective. But reason can also be used to *reflect* on our reasons, to bring them into accord with each other, and to modify them in the light of experience. In Adam Smith's view, self-reflective rules serve as a check on the unfettered expression of our moral intuitions. A commitment to behavior governed by moral rules helps us to avoid self-deception and acts as a brake on impulses we would later regret.

Reason may dictate fair behavior to be logically consistent. Sympathy has evolved the better to understand others, and reason the better to calculate costs and benefits of different options. In a situation in which I can benefit at your expense, I calculate my expected utility for each of my choices, and use sympathy to calculate your expected utility for each of your choices, the better for me to predict and counter your likely responses. The accuracy of my calculations depends on the quality of my simulation of your choices and the quality of my reasoning. Moreover, when I adopt a third-person perspective to anticipate the reactions of an observer, I elide the distinction between you and me. Why then, as a purely rational question, unconnected to emotion, should I value my utility higher than yours in choosing a course of action?

Emotion enters into my calculation of our respective utilities, and emotion may strongly advocate that my utility trumps your utility, but reason has considered the question from both sides and recognizes that the difference between *my* utility and *your* utility is an arbitrary criterion for breaking the fundamental symmetry of our situations. How often people choose their behavior based on such abstract considerations is an open question, but rational arguments of this kind are frequently used in attempts to persuade others of what they ought to do.

For Smith, the special prominence we give to our own interests is not arbitrary when viewed from the perspective of final causes: "Every man . . . is first and principally recommended to his own care; and every man is certainly, in every respect, fitter and abler to take care of himself than of any other person" (Smith 1976, 219). Man cares for himself because otherwise he would "have no motive for avoiding an accident which must necessarily diminish his utility both to himself and to society; and Nature, from her parental care of both, meant that he should anxiously avoid all such accidents" (148).

Cultural Elements

> Since our sentiments concerning beauty of every kind, are so much influenced by custom and fashion, it cannot be expected, that those, concerning the beauty of conduct should be entirely exempted from the dominion of those principles.
>
> —Adam Smith (1976)

Our bodily sympathy and direct observations of others, our emulation of their behavior and sympathy with their thought, our attempts to persuade and our changes of mind, our desire for approval and fear of rejection, our listening to friends and to strangers, to parents and teachers, to rabbis, ayatollahs, and priests, and our reading of texts and viewing of films have caused the clay of our moral intuitions to be molded by interpersonal interactions and cultural processes. Moral thought and moral practice are influenced by untold generations of argument and reasoning about moral dilemmas.

Within social groups, the frequencies of alternative moral concepts wax and wane as a consequence of backsliding, persuasion, conversion, and execution of heretics. New concepts are proposed and old concepts mutate. These concepts become organized into coherent moral codes that differ in the sentiments they reward and the sentiments they punish, in the forms of sympathy they encourage and the forms they discourage. The adherents of these codes, and those who live under a code's shadow, learn to modify their public selves accordingly. Conformity to explicit and implicit rules is a hallmark of most successful moral traditions, enforced by the fear of punishment and the fear of rejection. Among social groups, societies that were organized in ways that minimized internal conflicts and delivered more goods to their members possessed moral codes that were more likely to be emulated by other societies. And societies that were able to marshal military might to incorporate larger territories and larger populations under their moral code were more likely to have elements of their code copied by threatened neighbors.

A common feature of moral behavior is indignation against transgressors. Someone who is judged to have acted immorally has thereby placed himself outside the protection of the moral restraints we observe in our interactions with others. He deserves to be punished and made to feel pain. Many of the worst things done by humans to other humans have been perpetrated by groups who saw their actions as morally justified by the immoral behavior of their victims (Haig 2007a).

Moral codes are coercive. Their commands are considered universal and absolute by their adherents, and thus binding on all, regardless of personal preference. The "culture wars" of the contemporary United States can be viewed as an escalated conflict between alternative moral codes. If a policy can be defined as moral, much energy can be mobilized to advance it. If it can be defined as immoral, the forces of moral outrage can be marshaled against it. For these reasons, political debate is often framed in moral terms, as arguments about what is right, especially by those who pursue selfish ends.

Moral codes are the products of cultural evolution and have evolved self-protective adaptations. Certain forms of sympathy are prohibited or discouraged, especially sympathy toward competing moral principles. For many, it is a sin to experience sympathetically, nay even to think about the possibility of experiencing, a moral principle from a different perspective. These prohibitions apply both to the "politically correct" and the "culturally conservative." If I were to venture one suggestion as to how to move toward a truce on the battlefield of competing moral "absolutes" it would be to recognize, and then combat, our partisan prohibitions on sympathy. But both sides fear unreciprocated sympathy will be exploited. Politics becomes trapped in a prisoner's dilemma in which moral intransigence is the dominant strategy.

Responsibility

> When I endeavour to examine my own conduct, when I endeavour
> to pass sentence upon it, and either to approve or condemn it, it is
> evident that, in all such cases, I divide myself, as it were, into two
> persons; and that I, the examiner and judge, represent a different
> character from that other I, the person whose character is examined
> into and judged of.
> —Adam Smith (1976)

Many voices, both external and internal, tell me what I should
do. I am bombarded from without, by exhortations to do this or
do that, by threats of punishment and promises of reward, by
reasoned arguments and ecstatic visions; and I am beguiled from
within, by reason, conscience, duty, honor, hope, fear, contradictory passions, contradictory rules, and competing moral traditions that I have partially internalized. And, perhaps, there are
silent voices of an unconscious self, working behind the scenes
and responsible for my unaccountable impulses. Even different
genes within my genome may come down on different sides of
internal moral conflicts. But when a decision is made, "I" am
responsible, as the arbiter among the stakeholders of the self.
May the Lord have mercy on my soul.

9   How Come? What For? Why?

> It takes . . . a mind debauched by learning to carry the process of making the natural seem strange so far as to ask for the why of any instinctive human act.
> —William James (1887)

Adam Smith's account of the moral sentiments resonates with modern themes in evolutionary biology. His distinction between our reasons and the reasons for these reasons recalls evolutionary biologists' distinction between explanations of mechanisms and explanations of why those mechanisms have evolved. This distinction is commonly accredited to Ernst Mayr (1961) and his separation of proximate and ultimate causes, although I think this interpretation is a substantial misreading of Mayr's text. This chapter originated as my response to an invitation to comment on the distinction between proximate and ultimate causes. Whether this distinction clarified or obfuscated understanding of the evolutionary process had become a point of passionate contention in the philosophy of biology. Some material on pre-nineteenth-century uses of "proximate cause" and "ultimate cause" were not considered germane by the soliciting editor and were deleted at his request but have been restored in this version.

My purpose of including this chapter in the current volume is to illustrate that concepts of cause are not as straightforward as most biologists think, to illustrate the aversion many biologists have for teleological language, and to illustrate the confusions that can result from the ambiguities of words.

On the Remote Origins of Ultimate Causes

Ernst Mayr (1961) was a staunch advocate of historical explanations in biology: ultimate causes were "causes that have a history." He believed one could not fully understand biological processes without understanding their evolutionary history. A parallel argument can be made for the meanings of words. Meanings evolve by mutation, semantic drift, and competition among alternatives. At any one time, variant definitions may be adopted by different individuals, or by the same individual in different contexts, and the frequency of variants may change over time. Which senses prevail may shape the direction and outcome of scientific and philosophical debates. Protagonists want others to see the world in their terms. One benefit of addressing how meanings have evolved is to step back from arguments about the "true," "correct," or "real" meanings of words.

The distinction between proximate and ultimate causes in evolutionary biology has variously been interpreted as a distinction between causes in the present and causes in the evolutionary past or between explanations of mechanism and explanations of adaptive function. Everyone seems to agree what are proximate causes. These are Aristotelian efficient causes. Differences of opinion center around the understanding of ultimate causes. Ambiguity in the meaning of "ultimate cause" has existed for centuries. "Ultimate cause" could refer either to the first in a series of efficient causes or to a final cause, often conceived as coming at the end of a series of events. The current ambiguity—whether ultimate causes are historical explanations or functional explanations—can be considered a descendant of this older ambiguity in the meaning of ultimate cause. Current debates would be clarified by distinguishing between these different senses of ultimate cause.

The Oxford English Dictionary is a good place to begin. The adjectives *proximate* and *ultimate* are derived from Latin verbs for "to draw near" and "to be at the end." The primary definition of *proximate* is "Coming immediately before or after in a chain of causation. . . . Freq. in *proximate cause*. Opposed to *remote* or *ultimate*." The primary definition of ultimate is "Of ends, designs, etc.: Lying beyond all others; forming the final aim or object." The earliest use of both adjectives is dated to the mid-seventeenth century. Of particular significance, "ultimate cause" has long been associated with "final aims," and "proximate cause" has been variously opposed to "remote cause" or "ultimate cause."

Proxima causa and *ultima causa* were used in scholastic Latin for centuries before their adoption into English as *proximate* and *ultimate cause*. In *De principiis naturae*, Thomas of Aquinas presented an Aristotelian distinction between prior and posterior causes:

We must understand that *proximate cause* means the same as *posterior cause*, and *remote cause* the same as *prior cause*. Accordingly, these two divisions of causes—prior and posterior, and remote and proximate—signify the same thing. Furthermore, we should observe that the more universal cause is always called the remote cause, while the more particular cause is called the proximate. For example, the proximate form of man is his definition, that is, *rational mortal animal*; but animal is more remote, and substance even further removed. For all superior things are forms of inferior ones. Similarly, the proximate matter of a statue is bronze, while the remote matter is metal, and the more remote is body. (1965, 23)

Remote causes explain proximate causes, but not the reverse. In the above passage, Thomas illustrates the distinction between proximate and remote causes by using hierarchies of formal and material causes in which priority is generality: general causes are remote and specific causes proximate. However, the distinction also applies to chains of efficient causation where priority is temporal—earlier events cause later events—and to final causes where proximate ends are means to higher goals.

In any causal chain or hierarchy, the most remote cause, the cause without a prior cause, was the ultimate cause. In *Summa contra gentiles*, Thomas used Aristotelian arguments against infinite causal chains to demonstrate the necessity of ultimate causes. He reasoned that there must be a first unmoved mover in chains of efficient causation and a first cause that is an end in itself in chains of final causes (Aquinas 1975, chap. 13, 37–38). For Thomas, the unmoved mover and ultimate end (*ultima finis*) were one and the same. In the beginning is the end.

Spinoza similarly distinguished between proximate and remote causes. In his *Short Treatise on God* he averred

God is the *proximate* cause of the things that are infinite and immutable and which we assert to have been created immediately by him, but, in one sense he is the remote cause of all individual things. (2002, 51)

Proposition 28 of his *Ethics* argued for infinite chains of efficient causes in the realm of individual (finite) things. The Scholium to that proposition states:

It follows, firstly, that God is absolutely the proximate cause of things directly produced by him. . . . It follows, secondly, that God cannot properly be said to be the remote cause of individual things. . . . For by "remote cause" we understand a cause which is in no way conjoined with its effect. But all things that are, are in God, and depend on God in such a way that they can neither be nor be conceived without him. (2002, 51)

Spinoza seems to change his mind about whether God is a remote cause of individual things. In Proposition 28, God is eternally present, proximate and not remote. Spinoza categorically rejected final causes, "all final causes are figments of the human imagination" (2002, 59). Spinoza denied final causes even to God, because everything is in God and if God acted with an end in view he would necessarily be seeking something he lacked.

Proximate cause also has a rich legal history, especially in the law of torts. Francis Bacon's (1596) first maxim of the law was *In jure non remota causa sed proxima spectatur*. If every cause has prior causes, then the law pragmatically concerns itself only with the immediate or proximate cause rather than engage in an investigation of remote causes. In summary, causes were ordered by metaphoric or actual time (prior and posterior) or by metaphoric distance (proximate and remote). These axes were orthogonal to distinctions among efficient, material, formal and final causes. Ultimate causes could be invoked for all four Aristotelian categories of cause.

Proximate and Ultimate Causes in the Nineteenth Century

It is not my intention to review the diverse meanings and nuances of ultimate and proximate causes in the nineteenth century. I will limit myself to some illustrative examples from medicine and the philosophy of Herbert Spencer. In medicine, proximate (or immediate) causes of disease were distinguished from remote causes with occasional references to ultimate (or primary) causes. The second volume of Erasmus Darwin's *Zoono-mia* presented a classification of diseases according to proximate causes:

Thus in the cramp of the calves of the legs in diarrhœa, the increased sensorial power of association is the proximate cause; the preceding increased action of the bowels is the remote cause; and the proximate effect is the violent contractions of the musculi gastrocnemii; but the pain of these muscles is only an attendant symptom, or a remote effect. (1818, 361)

John Chapman similarly addressed the causes of diarrhea. It could arise

from the irritation of dentition, from irritating ingesta, from drinking impure water, from inhaling noxious gases, from ulceration of the bowels (as in phthisis), and the symptomatic diarrhœa of numerous diseases" but "however diverse the *primary* cause of the disease, the *proximate* cause is always the same . . . viz., hyperæmia of the spinal and sympathetic nervous centres. (1865, 14)

Here, the proximate cause was the final mechanism, elicited by diverse primary causes, that resulted in shared symptoms.

In the context of disease, ultimate causes were physical causes that occurred before proximate causes. Thus, in a discussion of potato blight, "the *Peronospora* is undoubtedly the proximate cause of the disease, for the ultimate cause we may look to a very different set of circumstances. . . . The attacks of the parasite may be favoured by special climatal or other conditions" ("The Potato Disease," 1872). Similarly, "a fall early in January last resulting in a fractured thigh was the ultimate cause of his death, which occurred on September 5" ("Rudolf Ludwig Karl Virchow: Obituary," 1902). By contrast, the "ultimate cause of death" in contemporary medicine is often the cause *at the end* of the series, immediately before death, rather than the remote precipitating cause *at the beginning* of the series. For example, "The ultimate cause of death in patients with sepsis is multiple organ failure. Typically, patients will first develop a single organ failure . . . and then if the disease remains unchecked, will progressively develop failure of other organ systems" (J. Cohen 2002).

Herbert Spencer described "the necessary antagonism of individuation and reproduction" as ensuring "the final attainment of the highest form of [racial] maintenance" and as leading to "the ultimate disappearance of the original excess of fertility" (1852, 501). He continued: "From the beginning, pressure of population has been the proximate cause of progress. . . . It compelled men to abandon predatory habits. . . . It forced men into the social state. . . . And after having caused, as it ultimately must, the due peopling of the globe, and the bringing of all of its habitable parts into the highest state of culture . . . after having done all this, we see that the pressure of population, as it gradually finishes its work, must gradually bring itself to an end." Proximate cause, in this passage, is the means that brings about a predestined or higher end. In a similar vein, the locomotive engine was "the proximate cause of our railway system [that] has changed the face of the country, the course of trade, and the habits of the people" (Spencer 1857, 481).

For George Stokes (1887), "the highest aim of physical science is, as far as may be possible, to refer phenomena to their proximate causes." However, beyond a certain point "Science conducts us to a void she cannot fill." Spencer had earlier addressed this void. In *First Principles*, he equated Ultimate Cause (capitalized) with the Unknowable "that through which all things exist" (Spencer 1867, 108–114). However, when he turned to the knowable, and the instability of homogenous bodies, certain slight differences among units were doubtless the "proximate cause" of heterogeneity but the "ultimate cause" (uncapitalized) was unequal exposure of the parts to incident forces (1867, 424).

In summary, proximate causes were physical causes. "Ultimate cause" could refer to a physical cause that occurred before a proximate cause or to a final cause. Finally, in preparation for my discussion of Mayr's distinction between proximate and ultimate causes, it is worth noting that efficient causes were often associated with *how* questions and final causes with *why* questions. I will give two examples from the nineteenth century of this contrast. First, Johann Peter Eckermann reports Goethe stating: "Die Frage nach dem Zweck, die Frage Warum? ist durchaus nicht wissenschaftlich. Etwas weiter aber kommt man mit der Frage Wie?" (1836, 283). In my imperfect translation: "The question of purpose, the question *why?* is not at all scientific. But one can get a little further with the question *how?*" Second, Charles Kingsley wrote in *First Course of Earth Lore for Children*: "But of one thing I must warn you, that you must not confound Madam How and Lady Why. Many people do it and fall into great mistakes thereby" (1873, 4). He identified Madam How with Nature and her actions with Fact (348). Her beautiful mistress, Lady Why remained unidentified "though she has a Master over her again—whose name I leave for you to guess" (3).

Ernst Mayr and Teleology

The attempts to answer the questions "Why" and "How"—"To what end?" and "In what way?"—by no means interfere with each other. . . . There are always these two questions to be answered with reference to any one natural phenomenon, and both must be answered if the facts are to be fully understood.

—Edward Poulton (1908)

Ernst Mayr presented his "Cause and Effect in Biology" to the
Hayden Colloquium on Scientific Method and Concept at MIT
during the academic year 1960–1961 (Beatty 1994; Lerner 1965).
A version was published in *Science* (Mayr 1961) and subsequently
in the proceedings of the Hayden Colloquium (Mayr 1965). The
two versions are highly similar but not identical. The 1961 version (which I will use in my discussion) is commonly presented
to students as distinguishing *how* questions (proximate causes)
from *why* questions (ultimate causes) with *why* interpreted as the
question of adaptive purpose, but this is a misinterpretation of
Mayr's text.

Mayr (1961) was not the first biologist to attach *how* to proximate causes and *why* to ultimate causes. A decade earlier, John
H. Mullahy had written:

Scientists as such are not prepared to answer the ultimate questions of
why there is a universe and *why* the universe evolves. The scientist is
only asked to describe the process; his profession demands only that he
tell us *how* the universe evolves. When scientists pause in their inquiry
into the proximate explanation of things and brashly hybridize their
professional status by taking the philosophical bit between their teeth,
the sterility of their effort serves only to remind us of other hybrids more
accustomed to the bit. (1951, 20)

Similarly, Claude Wardlaw wrote that biology could aspire to
an understanding of "the ultimate and proximate causes of evolution and of the mechanism of life and morphogenesis," but
the "reason for evolution, i.e. the *why* of evolution is generally
regarded as belonging to a branch of learning other than biology" (1952, 463). Here, Wardlaw used "ultimate cause" to refer
to remote physical causes. Like Mullahy (and before him Goethe
and Kinglsey), Wardlaw considered reasons *why* to lie outside
the domain of science. It should be noted that Wardlaw's and

Mullahy's *why* is the *what for* of the evolutionary process itself, not the *what for* of the adaptive products of evolution.

One must consider Mayr's "political" motivations to understand the goals of "Cause and Effect in Biology." An important motivation was to defend the place of systematics and evolutionary biology in the academy against triumphalist molecular biology (Beatty 1994; Dietrich 1998). The criticisms against which Mayr wished to defend his discipline are not difficult to divine, both from his text and because evolutionary biologists still encounter the same criticisms. Evolutionary biology, it is said, smuggles in unscientific, teleological concepts and is less rigorous and predictive than the "hard" sciences.

At the same time, Mayr wanted to disown any connection between evolutionary biology, correctly delimited, and various vitalistic theories that were current in his youth. His opening paragraph addresses the theories of Driesch, Bergson, and Lecompte du Noüy, and its last paragraph concludes: "The complexities of biological causality do not justify embracing non-scientific ideologies, such as vitalism or finalism" (1961, 1506). Given this prominence, one might be tempted to conclude that vitalism, rather than proximate and ultimate causes, was Mayr's principal target and that he wanted to drive a stake through the heart of theories that, at the time of his writing, were already feeble if not dead. His reason for flaying the dead horse of vitalism was tactical. Vitalistic theories had been an inevitable reaction to "Descartes's grossly mechanistic interpretation of life" (1501). Mayr wanted to argue that functional biologists' concepts of causation, derived from the physical sciences, were impoverished and inadequate for understanding the living world, whereas evolutionary biologists had a richer view; but he needed to defend himself against the accusation that he was invoking some *entelechy* or *elán vital* in addition to physical causation.

"Cause and Effect in Biology" is structured around three aspects of causality: explanation of past events, prediction of future events, and teleology. Mayr first distinguished two broad disciplines, functional and evolutionary biology, with different explanatory goals and different causal concepts. He identified the functional biologist's question as *how* and the evolutionary biologist's question as *why*. The functional biologist was concerned with immediate causes, whereas the evolutionary biologist was concerned with "causes that have a history." Thus, "proximate causes govern the responses of the individual (and his organs) to immediate factors of the environment, while ultimate causes are responsible for the evolution of the particular DNA code of information with which every individual of every species is endowed" (1503).

Because Mayr's "ultimate cause" is commonly interpreted as the explanation of selective value or purpose, it is worth quoting his explicit disavowal of this interpretation:

When we say "why" we must always be aware of the ambiguity of this term. It may mean "how come?," but it may also mean the finalistic "what for?" It is obvious that the evolutionary biologist has in mind the historical "how come?" when he asks "why?" (1502)

What did Mayr mean by "proximate" and "ultimate" causes and why did he choose these terms? More than once, Mayr states that proximate causes are immediate whereas ultimate causes are historical. Therefore, ultimate causes are presented as temporally prior to proximate causes. Mayr clearly did not believe that evolutionary causes were without prior cause, so why choose "ultimate" rather than, say, "remote"? This was politics. Proximate causes are more salient than remote causes, but *ultimate* valorizes the evolutionary over the merely *proximate*.

Mayr's (1961) treatment of teleology is revealing. He explictly disavowed "What for?" because of its teleological overtones. The idea that evolution had a goal or purpose was the feature of vitalistic theories to which Mayr most strongly objected. Goal-directed behaviors are present in biology, but these are caused by the playing out of evolved genetic programs. By implication, how these programs are expressed is the province of functional biologists, but the origin of these programs is the realm of evolutionary biologists. "The development or behavior of an individual is purposive, natural selection is definitely not" (1504). Mayr subtly shifts the stigma of teleological thinking from evolutionary to functional biology. Although evolution was not goal-directed, Mayr recognized that natural selection produced organisms with apparent purposes. Rather than describe goal-directed behaviors as teleological, he preferred to use the adjective *teleonomic* and to restrict this term "rigidly to systems operating on the basis of a program, a code of information" (1504).

Mayr (1969) returned to these themes in "Footnotes on the Philosophy of Biology." "Vitalism," he declared, "is now dead, as far as biologists are concerned. . . . The enormous complexity of biological systems, the historical nature of all organisms, and the fact that organisms contain a historically evolved genetic program make them so totally different from nonliving objects that generalizations derived from nonliving objects are in most cases meaningless or trivial when applied to phenomena of life. . . . Purposive processes and behaviors that owe the adaptive nature of their response to genetic programming, may be designated as *teleonomic*. That part of Aristotle's teleology is not acceptable according to which the universe as a whole is programmed to evolve harmoniously, ultimately resulting in a harmonious kosmos. This is teleology *sensu stricto* and for this no substantiation has been found in any area of science" (197–202).

Mayr's understanding of these issues evolved. Mayr (1974) discussed the relation between teleology and teleonomy: "*A teleonomic process or behavior is one which owes its goal-directedness to the operation of a program*" (98). The significant change in the reworded definition was that he deleted "system," which he now saw as static and not dynamic. He was now unwilling to concede an eye, or a torpedo before it is fired, as being teleonomic. "Goal-directed" was not the same as "purposive." "Teleonomy" was not synonymous with "adaptation." He had also shifted ground on the evolutionary *why. What* and *how* were now considered adequate for the physical sciences, but no biological explanation was complete without an answer to *why.* "It is necessary for the completion of causal analysis to ask for any feature, why it exists, that is what its function and role in the life of the particular organism is" (108). Asking *why* "demands asking for the selective significance of every aspect of the phenotype" (109). This is *what for?* rather than *how come?* Mayr (1992, 131) wrote: "Adaptedness is an *a posteriori* result rather than an *a priori* goal-seeking. For this reason the word teleological is misleading when applied to adapted features." Mayr (1993, 94) recognized the historical association of ultimate causes with natural theology and wrote: "In order to shed the historical impediment of the term ultimate, I have used in most of my recent papers the term 'evolutionary' instead of ultimate causation."

The older Mayr was notable for dogmatic, one might say *ex cathedra*, pronouncements. A personal anecdote may be of interest to historians of evolutionary biology. Early in 1999, Mayr summoned me to his office. "We are going to write a paper together," he declared. "My English friend," he hesitated—"John Maynard Smith," I suggested—"Yes," he confirmed, "is wrong: animals do not play games." This was just after the announcement that Mayr was to share the Crafoord Prize with John Maynard Smith and George Williams, so I asked him how he felt about sharing the prize with Maynard Smith. The prize was "well deserved," but not for Williams: "Everything he has ever done has been wrong!"

Proximate and Ultimate Causes after Mayr

Mayr's distinction between proximate and ultimate causes was widely adopted by evolutionary biologists, but largely ignored by functional biologists. Perhaps some of the attraction for evolutionary biologists was the connotations of "ultimate" and "proximate." It is easy to imply that ultimate causes are more important than proximate causes. But this invidious comparison certainly did not endear "ultimate" to functional biologists who were interested in proximate causes. "Ultimate cause" has been labeled "simply pretentious" (Francis 1990) and "disciplinary chauvinism" (Dewsbury 1999). Thirty years after Mayr (1961) first contrasted biological explanations in terms of proximate and ultimate causes, Beatty wrote that "more than minor disagreement with his insistence on the proximate/ultimate distinction would be heretical" (1994, 352). However, recent exchanges in the philosophy of biology suggest a doctrinal dispute has broken out, in which papal authority is challenged by latter-day Luthers (Laland et al. 2013b; Thierry 2005) and defended by contemporary Johns of the Cross (Dickins and Barton 2013; Gardner 2013).

These recent debates about whether the proximate–ultimate distinction is useful, outdated, or pernicious have involved much arguing at cross-purposes because proximate and ultimate causes have come to conflate two distinctions. The first is a distinction between immediate and historical causes. The second is a distinction between mechanism and adaptive function. Mayr (1961) emphasized the first distinction, but most evolutionary biologists who adopted his terminology emphasized the second. This has created a semantic morass in which some critics of Mayr's distinction employ a definition closer to Mayr's intended meaning than do *most* supporters of the distinction. Two criteria are often invoked to determine the "correct" definition of a term. One is historical: what did Mayr mean by "ultimate cause"? The other is majoritarian: what do most individuals in community x mean by "ultimate cause"? Unfortunately, these criteria suggest different answers to the "correct" meaning of "ultimate cause."

Ambiguity in "ultimate cause" arose from an ambiguity in *why?* that Tinbergen (1968) avoided by separating questions of survival value from evolutionary history. Mayr explicitly disavowed the finalistic *what for?* in favor of the historical *how come?* but many evolutionary biologists wished to defend *what for?* against the mechanistic *how? What for?* is exclusively concerned with natural selection, but *how come?* encompasses additional historical factors. Close attention to whether authors identify ultimate causes with *what for?* or *how come?* clarifies their positions and illuminates causes of misunderstanding in current debates. Some see the relevant distinction as predominantly temporal (of immediate versus historical causes):

This conceptual dichotomy is a deeply engrained habit of thinking and is characterized by the belief that aspects of development are determined by either (a) events which occurred earlier in the development of the individual, or (b) preontogenetic factors which operated on the ancestors of the individual. (Lickliter and Berry 1990, 349)

Proximate explanations focus on causes in the present; evolutionary explanations focus on how the present has been shaped by events in the past. (Hochman 2013, 593)

Proximate causes are immediate, mechanical influences on a trait. . . . Ultimate causes are historical explanations. (Laland et al. 2013a, 720)

From this perspective, the distinction between proximate and evolutionary causes is seen as a false dichotomy, without a principled distinction between *how* and *how come*. By contrast, defenders of a proximate–ultimate distinction usually focus on the difference between *how* and *what for* and implicitly or explicitly place *how come* on the proximate side of the ledger:

To understand behavior . . . it is vital to distinguish between the "proximate" cause (the physiological mechanism) and the "ultimate" cause (the evolutionary "goal") of the behavior in question. (Burnham and Johnson 2005, 124)

The explanatory utility of an ultimate account is not solely in a detailed phylogeny, capturing a particular trait's ancestry and shifting phenotype over time; it is the exposure of function, of why a given trait is as it is. . . . Historical accounts are not in any sense default ultimate accounts, they . . . can be understood in purely proximate terms. (Dickins and Barton 2013, 749)

Two issues have become enmeshed. The first concerns what efficient causes are important in evolutionary change. Do developmental mechanisms play a role? Does organismal evolution alter the environment and subsequent selection? Few would dissent from an affirmative answer to either question. The problem is that some define these as proximate and others as ultimate questions.

The second issue concerns the role of teleological reasoning and language in evolutionary biology. From this perspective, *how* and *why* separate questions of mechanism from questions of function. This separation is the evolutionary descendant of the venerable distinction between efficient and final causes, with final causes interpreted in a strictly Darwinian sense. Some are comfortable with Darwinized final causes, but others believe they have no place in science.

Cause-talk and function-talk are not simply different vocabularies, they are incommensurable. There is no common currency in which causal and functional explanations can be cashed in. There is no common currency for proximate and ultimate causes. There are no ultimate causes. (Francis 1990, 413)

The mechanisms behind evolutionary processes leading to novelty can be explained in the absence of ultimate (final) causes, but not in the absence of proximate (efficient) causes. (Guerrero-Bosagna 2012, 285)

"Why is A?" has two different, if related, meanings. If we are dealing with a conscious (or programmed) agent, we may ask why the agent took action A, seeking explanation of the agent's reason for action A. Absent such an agent, "why is A?" morphs into a "how?"-question from "how does A come about?" . . . Thus "why?" questions concerning natural selection actually are "how?" questions. (Watt 2013, 760)

Debate would be clarified by explicitly discussing mechanism versus function, or efficient versus final causes, rather than the ambiguous proximate versus ultimate. Part of the problem is disagreement about what constitutes a cause. Many would restrict "cause" to efficient causation and would deny that functional explanations are causal. Tinbergen (1968) belonged to this semantic camp. With respect to his four questions (survival value, mechanism, development, evolution), he wrote:

The first question, that of survival value, has to do with the *effects* of behavior; the other three are, each on a different time scale, concerned with its *causes*. (1412, emphases added)

However, a distinction between *how* (incorporating *how come*)? and *what for?* remains useful. Questions of mechanism and adaptive function invite different kinds of answers. Intentional language is often the most natural way to talk about adaptations. Accusations that such language illegitimately invokes supernatural entities or conscious agents are usually no more than willful misunderstanding and petty point-scoring.

In retrospect, Mayr's choice of "ultimate cause" to describe answers to *how come?* was unfortunate, because ultimate cause has an "affinity" for the finalistic *what for?* The reason is partly etymological. Ultimate refers to something "at the end" and final causes explain a thing by an "end." "Ultimate cause" and "final cause" are therefore easily perceived as synonyms. The reason is partly historical. The appearance of plan and purpose in Nature was commonly ascribed to God as the final or ultimate cause. Once appearances of purpose were explained as artifices of a blind (rather than divine) watchmaker, "ultimate cause" was readily interpreted as adaptive function. Mayr exacerbated the ambiguity by associating ultimate cause with *why* and proximate cause with *how*, a contrast that had historically been associated with the distinction between efficient causes *how?* and final causes *why?*

Modern defenders of a proximate–ultimate distinction interpret ultimate causes as answers to the teleological *what for?* They want recognition that "what is it for?" and "what is the mechanism?" are different questions with different kinds of answer and that both questions lie within the domain of science. They want respect for their interest in adaptive explanations. Modern critics of the proximate–ultimate distinction interpret ultimate causes as answers to the historical *how come?* They want recognition that answers to "what are the mechanisms?" are important for understanding "how did it evolve?" They want respect for their interest in the role of developmental mechanisms in evolutionary processes. Defenders and critics are talking past each other and, if they took a deep breath, would probably concede each other's principal concerns.

"Proximate cause" does not generate much fervor and remains useful, whereas "ultimate cause" arouses passions and means different things to the contending parties. Mayr (1961) explicitly disavowed *what for?* in favor of *how come?* Therefore, critics of the proximate–ultimate distinction can plausibly argue that their interpretation of ultimate cause is closer to Mayr's (1961) intent, whereas defenders of the distinction interpret the difference between ultimate and proximate causes as corresponding to the *why* versus *how* contrast of Mayr (1974). I recommend that we follow Mayr's (1993) lead and use "evolutionary cause" to refer to physical causes operating in evolutionary time, not just adaptation by natural selection. Proximate causes have acted in every generation in the past and can form a legitimate part of evolutionary explanation. On the other side of the contrast, those who care about distinguishing the functional *what for?* from the mechanistic *how?* should abandon "ultimate cause" and simply talk about purpose and mechanism or employ the venerable distinction between final and efficient causes, with final causes interpreted as ascriptions of function or anticipated ends. I do not know any vitalists among my colleagues, nor any colleague who believes that divine intervention plays a role in biological explanation. The presumption should be that when evolutionary biologists use teleological language they refer to adaptation by natural selection. There is plenty to argue about without squabbling over language.

10   Sameness and Difference

> One simply cannot escape the conclusion that the brain of a rat and
> a human are actually the "same" in spite of their obvious differences.
> —Günter P. Wagner (1989)

During the nineteenth century, naturalists and biologists sought
order in the patterns of sameness and difference observable
in the living world. William Swainson wrote, "The most ordinary observer perceives, that every created being has different
degrees of relationship or of resemblance to others. Where this
is immediate, it is termed affinity; where, on the other hand, it is
remote, it is a relation of analogy" (1835, 230). Attempts to clas-
sify the diversity of life became enmeshed with disputes about
the relation of structure to function. In comparisons among
species, structural similarities of organs with the same function
often appeared superficial relative to deep structural correspondences of organs with divergent functions. For some, this dis-
connect between functional and structural similarity indicated
that causes of structure existed independent of function. In
debates before the French Academy of Sciences in 1830, Georges
Cuvier argued that each organism was adapted to its conditions
of existence and that structure was shaped by functional needs.

In dissent, Étienne Geoffroy Saint-Hilaire argued that structural analogies among animals of very different outward form showed that bodily organization conformed to a common plan. Geoffroy saw his theory of analogues as avoiding the "seductive influence of forms and functions" (translation by Grene, in Le Guyader 2004, 111).

Richard Owen attended some of the Parisian debates and considered that Cuvier had the better argument. He was, at first, predisposed toward Cuvier's position, but his subsequent studies of the vertebrate skeleton caused a shift in allegiance toward belief in an archetype that represented the plan or idea on which all vertebrate skeletons were constructed (see Owen 1868, 787–789). Owen distinguished between homologous and analogous similarities. A homologue was "the same organ in different animals under every variety of form and function" whereas an analogue was "a part or organ in one animal which has the same function as another part or organ in another animal" (1848, 7). Although these definitions are often cited as the first clear statements of the concepts of homology and analogy, Owen himself wrote that in "illustrating the term *homology*, I have always felt and stated that I was merely making known the meaning of a term introduced into comparative anatomy long ago, and habitually used in the writings of the philosophical anatomists of Germany and France" (1846, 526). He saw his own conceptual contribution as distinguishing between general, special, and serial homology: general homology was the relation in which a part stood to "the ideal or fundamental type"; serial homology the relation between repeated parts within the same body; and special homology the "essential correspondence" of the same parts in different species.

Owen's archetypal plans undergirded outward form and function. Final causes were "barren and unproductive of the fruits
we are labouring to attain, and would yield us no clue to the
comprehension of that law of conformity of which we are in
quest" (1849, 40).

The "Bedeutung," or signification of a part in an animal body, may be
explained as the essential nature of such a part—as being the essential-
ity which it retains under every modification of size and form, and for
whatever office such modifications may adapt it. . . . [It signifies] that
essential character of a part which belongs to it in its relation to a predetermined pattern, answering to the "idea" of the Archetypal World in
the Platonic cosmogony, which archetype or primal pattern is the basis
supporting all the modification of such part for specific powers and actions in all animals possessing it. (1849, 30)

Owen's archetypes were descendants of Platonic ideas rather
than the Aristotelian union of formal and material causes. The
details of organic form were ascribed to secondary causes.

The recognition of an ideal Exemplar for the Vertebrated animals proves
that the knowledge of such a being as Man must have existed before
Man appeared. For the Divine mind which planned the Archetype also
foreknew all its modifications. The Archetypal idea was manifested in
the flesh, under divers such modifications, upon this planet, long prior
to the existence of those animal species that actually exemplify it. To
what natural laws or secondary causes the orderly succession and pro-
gression of such organic phænomena may have been committed we as
yet are ignorant. But if, without derogation of the Divine power, we
may conceive the existence of such ministers, and personify them by
the term "Nature," we learn from the past history of our globe that she
has advanced with slow and stately steps, guided by the archetypal light,
amidst the wreck of worlds, from the first embodiment of the Vertebrate
idea under its old Ichthyic vestment, until it became arrayed in the glorious garb of the Human form. (1849, 85–86)

In *Typical Forms and Special Ends in Creation*, James McCosh
and George Dickie (1856) attempted to reconcile typos and telos
in a revised natural theology that united the facts of homology
with the doctrine of final causes. They recognized a God of the
details in order and ornament as well as special adaptation:

We are inclined to admit that certainly in the vegetable, and probably in
the animal kingdom, there are parts retained for the sake of symmetry
which are not necessary for the mere function of the organ. In making
such an admission, we are not, so far as we can judge, weakening the
great principle of final cause, so long as we call in a higher final cause,
and affirm that these parts are fitted, in some cases, to give instruction to
mankind, and, in other cases, to gratify their higher tastes. (438)

The final cause of order was to render the world intelligible so
that man could put nature to practical use. The final cause of
ornament was to please our aesthetic sensibilities.

Miles Joseph Berkeley's (1857) *Introduction to Cryptogamic
Botany* illustrates how analogy and homology were understood
shortly before the publication of Darwin's theory of natural
selection: "Analogy . . . is always liable to seduce an inattentive
or ignorant observer into wrong notions as to the relation of
beings between which it exists," but "Homology is of far more
value; for when true it is founded on a deep knowledge of structure, and is indicative of either close or remote relation." Analogy
was "resemblance of *function*"; homology was "correspondence
of *structure* or origin." Homologous structures were "identical in
essence and origin" (39–41, emphases added; I interpret "origin"
as ontogenetic origin).

Darwin (1859) undercut appeals to Platonic ideas while nat-
uralizing final causes. He proposed that special homology was
explained by descent from a common ancestor: "If we suppose
that the ancient progenitor, the archetype as it may be called, had

its limbs constructed on the existing general pattern, for whatever purpose they served, we can at once perceive the plain signification of the homologous construction of the limbs throughout the whole class" (435). Moreover, with archetypes replaced by ancestors, general homology could be understood as special homology between present and past. Serial homology was jointly explained by correlations of growth and descent. "The several parts of the body which are homologous, and which, at an early embryonic period, are alike, seem liable to vary in an allied manner: we see this in the right and left sides of the body varying in the same manner; in the front and hind legs, and even in the jaws and limbs, varying together, for the lower jaw is believed to be homologous to the limbs" (143). Repeated parts were primordially similar but had diverged in form by natural selection to perform diverse functions. Nevertheless, "we need not wonder in discovering in such parts or organs, a certain degree of fundamental resemblance, retained by the strong principle of inheritance" (438).

The Origin of Species contributed to a general acceptance by morphologists of the *transformation* of species without a corresponding acceptance of Darwin's mechanism of natural selection. Owen was among many who decried Darwin's use of metaphor:

Assuming then that *Palæotherium* did ultimately become Equus, I gain no conception of the operation of the effective force by personifying as "Nature" the aggregate of beings which compose the universe, or the laws which govern these beings, by giving to my personification an attribute which can properly be predicated only of intelligence, and by saying, "Nature has selected the mid-hoof and rejected the others." . . . Such figurative language, I need not say, explains absolutely nothing. . . . One's surprise is that "tropes" and "personified acts" should not have died out, as explanatory devices, with the "archeus faber," the "nisus formativus," and other self-deceiving, world-beguiling simulacra of science, with the last century; and that a resuscitation should have had any success in the present. (1868, 794)

Toward a Genetic Concept of Homology

Questions of ontogeny and phylogeny were entangled in the minds of many biologists. During the nineteenth century, "evolution" could refer either to developmental change within a generation or transmutation of forms over many generations. After general acceptance that descent with modification could explain structural resemblance, E. Ray Lankester (1870) recommended that "homology" be discarded and replaced by two new terms: homogeny and homoplasy. He wrote, "Structures which are genetically related, in so far as they have a single representative in a common ancestor may be called *homogenous*" (36). By contrast, "When identical or nearly similar forces, or environments, act on two or more parts of an organism which are exactly or nearly alike, the resulting modifications of the various parts will be exactly or nearly alike. . . . I propose to call this type of agreement *homóplasis* or *homóplasy*" (39). Homoplasy included "all cases of close resemblance of form which are not traceable to homogeny, all *details* of resemblance not homogenous, in structures which are broadly homogenous, as well as in structures having no genetic affinity" (41). Homogeny depended "simply on the inheritance of a common part," whereas homoplasy depended "on a common action of evoking causes or moulding environment on such homogenous parts, or on parts which for other reasons offer a likeness of material to begin with" (42).

The difference between Darwin's and Lankester's explanations of serial homology is revealing. For Darwin, serial homology was explained by the liability of different parts of a single body "to vary in an allied manner" (1859, 143). For Lankester (1870), serial homology was homoplasy. Forelimbs and hindlimbs could not descend from the same part. Therefore, their concerted evolution must be explained by similar external forces acting on similar, but independently malleable, parts. At this stage of his career, Lankester believed in the direct inheritance of form with inheritance akin to memory (see his favorable review of Ewald Hering's and Ernst Haeckel's theories of heritable memory in Lankester 1876). A change to the forelimb of a progenitor could be "remembered" as a corresponding change in the forelimbs of descendants, but how could a change to an ancestral forelimb be "remembered" by subsequent hindlimbs? What was missing was Darwin's appeal to "internal" correlations of growth.

August Weismann (1890) recognized that contemporary enthusiasm for the inheritance of acquired characteristics was based on a model in which changes to parental parts were directly communicated to filial parts. He championed an alternative model, soon enthusiastically embraced by Lankester, in which determinants were inherited via nuclear chromatin but were expressed in cytoplasm during development. Serial homology and "correlations of growth" could now be explained by the expression of the same determinants in different locations within a single body. These determinants were conceptualized as genes in the years after the rediscovery of Mendel's experiments.

A concept of genetic homology arose from the theory that genes were physical structures with precise locations on chromosomes. Chromosomes of the same form paired at meiosis and were recognized as *homologous*. Genes at the same location on homologous chromosomes were called *alleles*. If alleles could trace their lineage back to a single material gene, then genetic homology could be rooted in common ancestry. Greater precision was given to the concept of genetic homology with the discovery of the double helix. Two DNA sequences were homologous if both were derived from a common ancestral template via unbroken chains of replication. Once genetic homology was defined as descent from the same template, homologous sequences need no longer exist at the same chromosomal locus and different parts of a sequence need not have the same ancestry. An ancestral sequence could leave descendants at multiple chromosomal locations by gene duplication and genome rearrangements (Fitch 1970), or cutting and splicing could create descendant sequences with parts that coalesced in different ancestral templates.

At first sight, genetic homology provides a means of identifying morphological homologs: two forms are homologous if their development is determined by homologous genes. However, the epigenetic nature of development means there can be no simple mapping of genotypic determinants to phenotypic forms. Body parts are generated anew in each generation by the interaction of many genes in environmental context (Waddington 1957). Although the question whether two DNA sequences are derived from a common template is relatively well-defined, the question whether two parts have evolved from the same ancestral part is problematic. "Characters are not literally derived from other characters. Organs are not descended from other organs, nor are they inherited from ancestors" (Cartmill 1994).

One resolution of this problem would be to reject talk of morphological homology as conceptually incoherent and restrict attributions of homology to DNA sequences. Such an approach would investigate the genetic architecture that underlies a character's development and ascertain how these networks and associated characters have been transformed in multiple lineages over evolutionary time. From this eliminationist perspective, the morphologist's question whether the phyllode of an *Acacia* is "really" a leaf or a petiole (Boke 1940) would be seen as wrong-headed. Rather, the developmental mechanisms that produce "phyllodes" would be dissected and the similarities and dissimilarities of this machinery with the mechanisms that produce "leaves" and "petioles" in related taxa would be described. The ontological question whether two parts belonged to the same *or* different kinds would lose meaning as more was learned about the mechanistic reasons for their similarities *and* dissimilarities. One could still talk of "leaves," "petioles," and "phyllodes" as terms of convenience without having an ontological commitment to leaves and petioles as natural kinds.

By contrast to problems with attributions of morphological homology, templated replication of nucleic acids provides a simple interpretation of claims that two sequences are derived from a common ancestral sequence. For many purposes, coalescence of gene sequences provides a perfectly adequate concept of homology. Nevertheless, practical situations will remain in which information about genetic mechanisms is unavailable but one would like to make statements about common "ancestry" of morphological features. One could then employ concepts of morphological homology as heuristic tools without a metaphysical commitment to homologs as natural kinds.

Characters and States

The problem with the definition of homologs as "the same organ under every variety of form" was that no unambiguous criteria were provided for deciding when organs of different form were the "same organ" and when they were "different organs." This problem was not resolved by replacing archetypes with ancestors. Nor was it resolved by replacing belief in the direct inheritance of form with belief in the inheritance of determinants of form, especially when determinants were identified by their effects on form. Wagner's (2014) *Homology, Genes, and Evolutionary Innovation* is a bold attempt to rescue morphological homologues as natural kinds. I was invited to write an extended book review, on which this chapter is based. My review attempted to find areas of consensus between functionalist heirs of Cuvier and structuralist heirs of Geoffroy, to identify where different perspectives of the same scene told the same story in different words.

Wagner diagnosed a conflict between structuralist and functionalist styles of thinking. Functionalists explain organismal traits by their adaptive value, whereas structuralists explain why things are the way they are by appeal to structural constraints and capacities. Wagner proposes "to overcome this conflict by addressing a specific biological phenomenon for which the conflict often crystallizes: the question of homology. . . . At its core, the question is whether homologs exist—that is whether they are natural members of the 'furniture of the world' or whether they are only transient traces of the phylogenetic past. In the latter case, they would have no biological, conceptual, or causal significance. In the former case, homologs would have to play a central role among the concepts of evolutionary theory" (2014, 8). Wagner opts for the central importance of homologs.

For Wagner, "the realization that complex organisms/systems have unique and historically contingent variational constraints and biases paves the way for a seamless unification of functionalist and structuralist agenda" (19). In Wagner's synthesis, conserved structural properties have a causal role in determining how structures vary, and fail to vary, over evolutionary time. Although Wagner offers an olive branch of unification, his book is clearly written to correct the myopia and astigmatism of adaptationists who, Wagner believes, have belittled, misrepresented, and misunderstood structuralists. Rapprochement should occur on structuralist terms (as befits the injured party). A functionalist who describes similar phenomena in functionalist language is likely to feel misrepresented and misunderstood and to insist that rapprochement occur on functionalist terms. And thus, an underlying consensus may be obscured by semantic arguments because human nature is quicker to recognize when we are misunderstood than when we have misunderstood.

Wagner advocates investigation of how the gene networks underlying character development have evolved, but his purpose is not to supplant the classical concept of morphological homology with one based on genetic homology. Rather, he seeks to identify the aspects of genetic control that form the mechanistic basis of character identity. Homology, he writes, "reflects the developmental organization of organisms" (2014, 72). The division of bodies into homologs carves an organism at its joints, sometimes literally. In his view, body parts exhibit evolutionary continuity that transcends changes in most aspects of the genetic control of their development.

Wagner distinguishes character *identities* from character *states*. Each identity can exist in multiple states. He sees the relationship between identities and states of morphological characters as "the same as that for *gene identity* and *alleles* in genetics" (2014, 54). Character identity corresponds to the "same organ" and character state to "every variety of form and function" in Owen's definition of homologs. Wagner defines *novelty* as the origin of new character identities and *adaptation* as evolutionary modification of character states. Wagner's model of character development recognizes three tiers of genetic control. The first tier is responsible for positional cues and the third for the realization of character state. The crucial middle tier is the Character Identity Network (ChIN). ChINs are mutually exclusive functional units that are more strongly conserved than the other tiers of genetic control, thus creating continuity of character identity, and quasi-autonomy from other characters, despite changes of character state and location. He proposes that the genetic changes that create novelty, and a new ChIN, differ in kind from those that produce adaptation by the modification of existing ChINs. This prepares the ground for his contentious claim that natural selection can explain adaptation but cannot explain novelty.

The lens of vertebrate eyes exemplifies a body part that can be homologized across species. In Wagner's terminology, the lens is a character identity that has evolved different states in different species. Wagner uses regeneration of the lens in newts to illustrate how homologous organs need not have the same developmental origin nor follow the same ontogenetic pathway. The lens of a newt usually develops from ectodermal cells of the embryonic eye cup, but if the lens is experimentally excised, a new lens develops from marginal cells of the iris (2014, 84). One might say that the old and regenerated lenses are serially homologous in time.

Mature lens cells lack nuclei and consist of 30 to 50 percent protein in aqueous solution. The most abundant lens proteins are called crystallins and function in the refraction of light (Graw 2009). Different enzymes have been recruited as crystallins in different taxa (Piatigorsky 2007; Wistow 1993). For example, lactate dehydrogenase A (LDH-A) has been recruited as the υ-crystallin of the platypus (van Rheede et al. 2003), whereas lactate dehydrogenase B (LDH-B) has been recruited as the ε-crystallin of crocodiles and birds (Brunekeef et al. 1996; Wistow, Mulders, and de Jong 1987). The genes that encode LDH-A and LDH-B diverged from each other after a whole-genome duplication that occurred early in vertebrate evolution (Stock et al. 1997). The principal requirement to serve as a crystallin is that a protein remain in solution at high concentration to maintain lens transparency without aggregation to form cataracts. This property must be exceptionally stable because lenses must function for the life of an animal without protein turnover. Conserved properties of lactate dehydrogenases, such as solubility and compact structure, probably predisposed LDH-A and LDH-B for "independent" cooption as crystallins in monotremes and archosaurs.

Can a principled distinction be maintained between characters and states? The presence of ε- and υ-crystallins are character states of the lens, but does an enzyme's cooption as a crystallin also create a new character? Is this a minor regulatory change, a mere tweaking of expression levels in the lens, or a novelty, an enzyme doing double duty as a structural protein? There are many features of lenses about which one could make meaningful statements of shared or nonshared ancestry. Wagner (2014, 73) sees the failure of the historical concept of homology to clearly distinguish between characters and their states as a weakness because it replaces a notion of sameness with one of residual similarity, but a rigid distinction between characters and states risks restricting attributions of homology to individualized parts with chiseled ChINs.

Natural and Nominal Kinds

Wagner recognizes a continuum between differences of character state and differences of character identity, but he denies that this blurred boundary undermines his advocacy of a state–identity distinction because things toward one end of the continuum are clearly character states whereas things toward the other end are clearly character identities (2014, 198). More generally, he recognizes a continuum of conservation from the relatively invariant to the evolutionarily ephemeral, but champions a philosophical position that highly conserved attributes are sufficiently stable to be defining properties of natural kinds.

Wagner's major reason for considering biological kinds to be classes rather than evolutionary individuals is that homologs and taxa can possess invariant properties. Thus, "the 'essence' of being a eukaryote resides in the manner in which their cells are organized and how their genetic material is packaged" (2014, 236). But this essence of a natural kind excludes most dinoflagellates from the ranks of eukaryotes. Most eukaryotic DNA is "packaged" into nucleosomes, formed by coiling of the double helix around a complex of histone proteins, but dinoflagellate DNA is organized as permanently condensed, liquid crystalline chromosomes that lack histone proteins. Other morphological and molecular characters clearly indicate that dinoflagellates evolved from within the alveolate clade of eukaryotes (Gornik et al. 2012). One could undoubtedly modify the "defining characters" of eukaryotes to include dinoflagellates, but this new definition would be vulnerable to the next exception that probed the rule. It seems simpler to define eukaryotes as parts of a clade, an evolutionary "individual" with a unique evolutionary history, rather than as a class defined by essential characters.

Consider vertebrate kidneys. Kidneys develop from nephrogenic cords of mesoderm. The embryonic kidney, known as a *pronephros*, differentiates from the cranial end of a nephrogenic cord. A *mesonephros* subsequently develops from more caudal regions of the cord and replaces the pronephros as it degenerates. The adult kidney of fish and amphibians is considered to be a mesonephros, but the mesonephros is replaced as the functional adult kidney of reptiles and mammals by a *metanephros* that develops from the far caudal cord, although parts of the mesonephros persist as ducts of the male epididymis. Neither fish nor amphibians possess a metanephros (Hamilton, Boyd, and Mossman 1947).

The successive kidneys of reptiles and mammals—pronephros, mesonephros, metanephros—are serial homologs. Mesonephric kidneys of amphibians and epididymal ducts of mammals are special homologs. How can one define kidneys as a class of which all members are mutual homologs but no nonmembers are homologs? Is the epididymis a kidney in disguise, or do epididymides and kidneys possess distinct character identities? Do the mesonephric kidneys of amphibians and metanephric kidneys of mammals belong to the same or different classes? Answers could be found for each of these questions that would rescue "kidney" or "metanephros" or "epididymis" as natural kinds, at least temporarily, but such answers must be sought anew for each and every homologous part, and such ad hoc solutions must be revised each time something is found that does not neatly fit an existing definition.

Wagner (2014) prefers precise definitions to fuzzy concepts and rejects the idea that homologs are nominal kinds, but his depiction of nominal kinds as "simple arbitrary summaries" and "human-made distinctions for our own convenience . . . otherwise meaningless" (229–230) stacks the deck against nominalism. A nominalist could counter that all categories are human constructs but that some categories are less arbitrary, and more useful, than others. Indeed, a consistent nominalist would not object to useful categories being labeled natural kinds because she would accept that "natural kind" is itself a nominal kind that represents whatever a linguistic community chooses it to mean.

Wagner disarmingly acknowledges that nothing in his book should be construed as a formal definition. Rather, he sees himself as presenting models that will evolve into precise definitions as more is learned about developmental mechanisms (2014, 242–244). The opposite may be true. As more is learned, more forms may sit uncomfortably with respect to any definition one may care to propose. The nominalist would concede that homology concepts are useful tools but would affirm that different tools are suited to different tasks. In her view of conceptual pluralism, different disciplines will inevitably adopt different homology concepts, just as different disciplines employ different species concepts.

The identification of homologous parts in disparate organisms is often presented as conclusive evidence for their evolutionary descent from a common ancestor, but this commonplace argument hides an ontological tension. Evolutionary thought is inimical to rigorous definitions and well-defined categories because it concerns processes by which things of one kind become things of other kinds, whereas attributions of homology attempt to capture that which remains unchanged despite transformations of form and function. The more things change, the more they remain the same.

Novelty and Adaptation

Wagner argues that the "origin of homologs" is distinct from "their modification by natural selection" (2014, 43). He writes, "It is conceptually . . . necessary to distinguish between the evolution of adaptations and the origin of novelties" because "there are a number of features characteristic of novelties that make it unlikely that the adaptationist program will give us satisfactory answers" (121–123). In particular, the genetic rewiring of regulatory networks that creates novelties differs in kind from the mere tweaking of existing pathways that produces adaptation. As a consequence, "innovation is a different kind of process than is adaptation, which is usually studied within populations at the micro-evolutionary level" (209). Novelties are both rare and pregnant with possibilities.

There is more than one way to carve evolutionary biology at the joints. Wagner equates adaptationism with the micro-evolutionary level of population genetics (2014, 10–12), but many population geneticists would deny the adaptationist label, and many adaptationists, such as myself, would agree with Wagner that the genetic variation currently segregating in populations is not typical of the genetic changes that were responsible for major evolutionary innovations. Ilan Eshel (1996) has argued that the short-term models of population genetics employ different equilibrium concepts from models of long-term adaptation and that the postulate that "the behaviour of the long-term process can be fully understood by extrapolation of the analytically well-defined short-term process . . . is mathematically wrong." I believe that Wagner is misdirecting his disagreement with population genetics by labeling their models "the adaptationist program."

Wagner parts company with adaptationists when he down-
plays the role of adaptation in the genesis of novelty: "The
specific new potential of a novelty can hardly be 'seen' by natural selection that originally selected the new trait"; and "it is
unlikely that natural selection can provide a satisfactory account
of the fact that feathers turned out to be able to support flight,
whereas hair did not" (2014, 123). But no adaptationist would
ascribe foresight to natural selection. The evolutionary potential
of an innovation is always recognized in hindsight, with some
changes judged retrospectively more significant than others.

Wagner ascribes significant responsibility for determining the
direction of evolutionary change to sources of variation rather
than assigning the sole directive role to the sifting of this variation by natural selection. In particular, he proposes that rewiring of genetic networks occurs not by natural selection of point
mutations but by cooption of ready-made promoters from transposable elements that are only episodically active and lineage
specific. "Evidence that transposable elements play a major role
in the evolution of gene regulatory networks affects various uni-
formitarian ideas that are broadly accepted in evolutionary biology. . . . It is not far-fetched then to think that the evolutionary
fate of a lineage is strongly influenced and different from that of
other lineages, in part, because of the nature of genomic parasites that infect its genome at any point in time" (2014, 207).

Most adaptationists would include insertions of transposable
elements under their broad definition of "mutation" and would
emphasize that most insertions, like most point mutations, are
deleterious or selectively neutral. Deleterious insertions are eliminated by "negative" selection whereas neutral insertions either
drift to extinction or are eventually degraded by mutation.
Only a small minority of insertions are preserved by "positive"

selection, namely, those that serendipitously enhance adaptation (Haig 2012, 2016).

The analogy of electrical circuits may be helpful. Transistors and integrated circuits have revolutionized what is technologically possible, but not all circuits can be readily reconfigured as new devices. Wagner suggests adaptationists overemphasize the role of unconstrained natural selection and pay insufficient attention to the properties of components and existing circuits. Adaptationists argue that components do not self-assemble into novel gadgets without an electrical engineer. They emphasize the role of the engineer (natural selection), both in assembly of circuits and origin of components.

Consider the expression of prolactin (*PRL*) in the endometrium of elephants, rodents, and primates but not in the endometrium of rabbits, pigs, dogs, armadillos, or opossums (Emera et al. 2012). Three independent origins of endometrial *PRL* expression are associated with insertions from four families of transposable elements that occurred at different times during mammalian evolution (Emera et al. 2012; Lynch et al. 2008). Some insertions were not associated with endometrial expression for many millions of years (Emera and Wagner 2012a). All these insertions survived the sieve of natural selection, but there were undoubtedly many more insertions in *PRL* genes that have not left descendants or detectable traces in extant genomes. Are we to understand that endometrial expression would be adaptive in all mammals but an endometrial promoter never arose in the lineages of rabbits, pigs, dogs, or armadillos because of the absence of the right kind of transposable element? Or should we conclude that endometrial expression has arisen many times but has only been retained by natural selection in three lineages? Are transposable elements responsible for the origin of novelty, or is the creative agency the winnowing of insertions by natural selection? Do transposable elements rewire regulatory networks, or does natural selection rewire networks using the promoters of transposable elements as handy components?

It has been suggested that transposable elements (including retroviruses) have rewired endometrial and placental regulatory networks (Chuong 2013; Chuong et al. 2013; C. J. Cohen, Lock, and Mager 2009; Emera and Wagner 2012b; Lynch et al. 2011). Although insertions of transposable elements can occur almost anywhere in the genome, including in crystallin genes (Nag et al. 2007), their promoters have never been reported to confer lens-specific expression, or rewire regulatory networks of the lens, despite the recruitment of diverse genes as crystallins during vertebrate evolution (Wistow 1993). Is there something special about the endometrium and the placenta? Retroviruses are probably a source of placenta-specific and endometrium-specific promoters because retroviral expression in these tissues facilitates infectious transmission between mother and child (Haig 2012, 2013), but retroviruses are not expected to possess lens-specific promoters because replication in lenses does not facilitate retroviral transmission.

The endometrium and placenta appear to evolve more rapidly than the lens. Wagner would assign much of the credit for accelerated evolution at the maternal-fetal interface to transposable elements as a source of saltatory genetic variation. An adaptationist would explain rapid evolution of tissues that separate mother and fetus as an outcome of antagonistic selection between genes expressed in the maternal endometrium and genes expressed in the fetal placenta; between genes of maternal and paternal origin in the placenta; and between retroviral adaptations and the defenses of maternal and filial hosts (Haig 1993b, 2008b, 2012).

In one of his many beautiful metaphors, Darwin addressed the relation between selection and sources of variability:

> I have spoken of selection as the paramount power, yet its action absolutely depends on what we in our ignorance call spontaneous or accidental variability. Let an architect be compelled to build an edifice with uncut stones, fallen from a precipice. . . . If our architect succeeded in rearing a noble edifice . . . we should admire his skill even in a higher degree than if he had used stones shaped for the purpose. So it is with selection, whether applied by man or by nature; for although variability is indispensably necessary, yet, when we look at some highly complex and excellently adapted organism, variability sinks to a quite subordinate position in importance with selection, in the same manner as the shape of each fragment used by our supposed architect is unimportant in comparison with his skill. (1883/1998, 236)

The promoters that transposable elements disperse throughout the genome are a source of ready-made functional elements, shaped by previous natural selection. It is as if Darwin's architect were choosing stones from the rubble of Rome rather than from the scree at the foot of a precipice. The architect's task would undoubtedly be made easier by a source of previously shaped stones. This is an example of the reuse of old parts in the assembly of a novel gadget, which Darwin had previously illustrated with his metaphor of old springs, wheels, and pulleys, slightly altered. If I may express a pet peeve, it is with the current popularity of the neologism "exaptation" (use of something evolved for one function for another function) with the implication that the word identifies a previously unrecognized evolutionary principle distinct from adaptation.

Modularity and Evolvability

A genome's nucleic acid sequence can be likened to the software of a robotic control system. Among the important tasks it must control is the assembly of its own hardware. Such metaphors, of organisms as robots and genomes as software, are often dismissed because they devalue organismal autonomy and privilege genes over environment. But the objections seem overstated. Useful robots regularly make autonomous decisions and adjust their behavior in response to environmental inputs. Comparisons between genetic and robotic control systems are useful, both for the similarities and dissimilarities revealed.

Software engineering encompasses synchronic goals—writing software that is useful now—and diachronic goals—writing software that will be easy to modify in the future. With respect to future modification, software should be robust, so that changes do not break what already works, and open-ended, so that new functions can be integrated with minimal change (Calcott 2014). Software development is usually decomposed into manageable parts that can be independently programmed with a separate interface that calls upon modules as needed. Modular design has synchronic benefits—division of labor (separate teams can work simultaneously on distinct tasks without needing to constantly communicate) and comprehensibility (system function can be studied one module at a time)—overlapping with diachronic benefits—reduced pleiotropy (changes within a module do not ramify to other parts of the program) and reuse (self-contained modules can be adapted for novel functions) (Calcott 2014; Parnas 1972).

The diachronic benefits of modular software probably translate to genetic evolution, but the synchronic benefits may not.

In software engineering, each module is tested and debugged before interactions among modules and global functioning are tested. When the assembled system does not perform as intended, modularity facilitates the isolation and correction of problems (trouble-shooting). Natural selection does not divide labor among teams, cannot comprehend the code that it generates, and does not isolate problems before correcting them. Modules (if these exist) are tested not individually but as an ensemble. Negative selection to maintain existing genetic code is costly. An organism must die without progeny to eliminate a bug in any part of the system. These costs of negative selection create an advantage for using the same code for multiple functions, because a lethal failure to perform one function purifies shared code for all functions.

Modularity and evolvability can be designed features of software, but can they also be evolved properties of genetic systems? Programmers can anticipate future needs, but natural selection lacks foresight. For Lynch, "there is no compelling empirical or theoretical evidence that complexity, redundancy or other features of genetic pathways are promoted by natural selection" (2007, 810), whereas for Calcott, "complex integrated systems, whether evolved or engineered, share structural properties that affect how easily they can be modified to change what they do" (2014, 293). Lynch sees most proposals about the evolution of evolvability as adaptationist overreach, whereas Calcott sees evolvability as distinct from adaptation. Evolutionary arguments about evolvability are a semantic morass (Sniegowski and Murphy 2006).

When algorithms evolve to solve a task rather than are designed to solve the task, the evolved algorithms provide mixed support for the idea that natural selection favors modular architectures. Nonmodular algorithms tend to outperform modular algorithms when given a unitary task. Modularity, if initially present, tends to break down, because there are many possible connections that break modularity and increase fitness. On the other hand, algorithms spontaneously evolve modularity if given multiple tasks that alternate in a regular fashion (Kashtan and Alon 2005).

If the source code of a successful software package were examined, some parts of the code would be highly conserved and others would have undergone extensive change since the earliest versions. Some modules might have entirely changed their code while maintaining conserved function and conserved links to other modules. By contrast, if the evolution of a once successful software package were examined, the source code would be found to have initially undergone updating and addition of new functions, then updates would have slowed and eventually ceased, although the package would have continued to be used by an ever-dwindling number of users until its eventual extinction. There are many reasons why packages become "extinct," but one factor could be structural features of the source code that were not conducive to efficient updating and modification for new uses. By such a process, one would observe preferential survival of more evolvable software.

Every genome encodes features that are anciently conserved because changes are not compatible with a viable organism. Some of these developmental constraints may be more or less conducive to changes in other features that allow adaptation to changing environments. Thus differential extinction will result in the preferential survival of lineages with more evolvable genomes. This selection among lineages can be considered to choose among developmental constraints. Over evolutionary time, "good" constraints that promote long-term survival in a changing world, or prevent a lineage falling into short-term evolutionary traps, will tend to be preserved, whereas "bad" constraints that impede adaptive change will be eliminated.

I believe consensus exists between structuralists and adaptationists about what might be called "clade selection" of evolvability or evolutionary constraints where the latter are conserved features that influence evolutionary outcomes. However, clade selection is not sufficient. Conserved features must be maintained by negative selection within populations. One can ask whether loss of evolvability is one of the reasons for a feature's maintenance by negative selection or whether effects on evolvability are incidental by-products of negative selection for individual benefits. In general, mechanisms maintained by immediate benefits to organisms will be more robust than mechanisms maintained by effects on evolvability. One could also ask about the reasons for the origin of constraints by positive selection and whether there are reasons why positive selection for immediate benefits should favor mechanisms that have enhanced evolvability as a by-product.

Hypoxic cells of vertebrates release signals that trigger nearby blood vessels to grow toward the hypoxic region, alleviating hypoxia. This mechanism enhances evolvability because body parts are automatically supplied with blood vessels as they evolve new shapes (Gerhart and Kirschner 2009), but the mechanism is maintained by individual-level selection because organisms in which it malfunctions experience immediate costs. The benefit of enhanced evolvability for the lineage is an incidental by-product of the individual-level benefit. Perhaps plastic mechanisms, as exemplified by the vascular response to hypoxia, are superior to rigid mechanisms because plastic mechanisms allow adaptive responses to environmental or within-population variation. Thus, facultative responses favored by individual benefits may also enhance evolvability.

One can retrospectively conclude that morphologically diverse clades have had more evolvable genomes than extinct clades or "living fossils." Many will be tempted to extrapolate that clades that have been evolvable in the past will be evolvable in the future, but past extinctions of previously dominant clades should temper enthusiasm for this prediction. One can more confidently predict that some lineages that are now judged to have been evolvable will become extinct.

Formal Causes

Selection chooses from a set of alternatives. In the formalism of the strategic gene, natural selection of "allelic" differences requires three components of choice: a genetic difference, a phenotypic difference, and a selective environment. In this formalism, phenotype contains all things that differ, and environment all things that are the same, for the items of choice. Thus conserved features of bodies and genomes are part of the selective environment that chooses among genetic alternatives based on their phenotypes. The genetic variant chosen constitutes a record of the choice. When the choices of the environment are consistent and repeated, then one of the alternatives can become a fixed part of the selective environment for other choices. By this means, natural selection converts the variable (that which is selected) into the invariant (that which selects).

Processes by which genetic differences cause phenotypic differences can be considered efficient causes, and those by which phenotypic differences cause differential genetic replication can be considered final causes. Such processes take place in evolved structures (let us call them bodies) that consist of physical stuff (material cause) and inherited information about past choices (formal cause) of which the genome is the "textual" record. For each particular choice, the selective environment includes all aspects of bodies and genomes shared by the alternatives.

Genomic features are highly conserved if most genetic differences (mutations) have phenotypic effects that cause elimination of the difference (negative selection). These "invariant" features of the genome, and the bodily forms they determine, are part of the selective environment that chooses among differences in less conserved parts of the genome. Conserved features of the body and genome are often more conservative than the "external" environment and form the mechanistic basis of unity of type. These features confirm "the structuralist intuition that complex systems play a causal role in determining their evolutionary fates" (Wagner 2014, 18).

Vertebrates with two pairs of lateral appendages have moved from water, onto land, into air, back to water many times, and share many of their habitats with six-legged insects and eight-legged spiders. The original reasons why an ancestral vertebrate evolved two pairs of fins, whereas an ancestral insect evolved six pairs of legs and an ancestral spider eight pairs of legs, are probably lost in the depths of time. And these "original" reasons are distinct from the reasons why these numbers have been maintained ever since by negative selection. Paired pectoral and pelvic appendages have been among the most highly conserved elements of the selective environment in which the adaptive radiation of vertebrates has taken place. The body plans (*Baupläne*) of vertebrates, insects, and spiders can be judged retrospectively to have been stable platforms for evolutionary discovery.

Evolution is a recursive process that dissolves traditional distinctions between cause and effect when considering relations
between kinds (type causation) rather than between instances
of those kinds (token causation). Genes have a causal role in the
production of bodies, and bodies have a causal role in determining which genes survive the filter of natural selection. Form is
shaped by genetic networks, but form may persist while those
networks are radically refashioned under selective constraints
of form. Where does continuity of form reside in the flux of
efficient causes? Wagner (2014) would locate that continuity in
Character Identity Networks (ChINs). These "are the most conserved part of the gene regulatory network that underlies character development and, thus, are most consistently associated
with manifest character identity" (186). But, given enough time,
could not a ChIN be changed beyond recognition and yet a character remain "the same"?

Capsid proteins of diverse viruses exhibit structural similarities that are unlikely to be explained by convergence despite an
absence of detectable similarity in amino acid sequence (Bamford 2003). These proteins are encoded by DNA sequences that
descend from an ancestral sequence that encoded a capsid protein more than a billion years ago. Although structural similarities suggest that capsid proteins are genetically homologous by
the criterion of descent of their genes from a common template,
no similarity can be detected in either the nucleic acid or amino
acid sequence. We have returned full-circle to comparative
morphology. Common ancestry is suggested by shared possession of "double-barrel trimers" and "jelly rolls" recognized by
three-dimensional gestalt rather than linear sequence (Bamford,
Grimes, and Stuart 2008; Benson et al. 2004). Form has dictated
what has been conserved. The story has been retold so many

times with embellishments, abbreviations, and subtle rewording that evidence of common ancestry is no longer detectable in the text but confined to structural features of the plot.

Asa Gray (1874) pronounced the marriage of morphology and teleology but the couple have continued to argue, each with a grievance that they were not being heard by their spouse. Darwin wanted to give the last word to teleology, "the law of the Conditions of Existence is the higher law; as it includes, through the inheritance of former adaptations, that of Unity of Type" (1859, 206). But morphologists muttered under their breath that "Unity of Type is the higher law because the Conditions of Existence can only fashion new adaptations by modification of pre-existing forms." The argument continues because humans have great difficulty thinking and talking about recursive processes, witness Kant: "For a body therefore which is to be judged as a natural end in itself and in accordance with its internal possibility, it is required that its parts reciprocally produce each other, as far as both their form and combination is concerned, and thus produce a whole out of their own causality" (2000, 245). This chapter can be considered a continuation of this spousal squabble between Günter Wagner and myself, each of us offering reconciliation on our own terms. Wagner interprets living things through structuralist lenses. I interpret them through functionalist lenses. If our crystallins are free of cataracts, shouldn't we agree on what we see? No, because perception is more complex than projection of an image on a retinal screen, and there are many layers of interpretation between light entering the eyes and words appearing on paper.

Deeply conserved structures are indeed significant parts of the living world and have had profound influences on the course of evolution. Organisms and their genomes are mosaics of features of different evolutionary age. Older features, maintained by negative selection, comprise part of the selective environment that shapes the evolution of newer features by positive selection. Body plans and body parts are among the most conserved elements of the selective environment in which new mutations are tested. Well-trodden paths of development direct and constrain paths of evolutionary change. Form plays a selective role in the molding of form. These are all conclusions an adaptationist should be willing to concede without having to renounce natural selection as an explanation of conserved structures. Structuralism and functionalism can both be vindicated.

Form fascinates when one no longer has the force to understand force from within itself. That is, to create. This is why literary criticism is structuralist in every age, in its essence and destiny. (he who shall not be named)

11 Fighting the Good Cause

Many four-year-olds delight in asking "Why?" and when one offers an explanation, respond with another "Why?" requesting explanations of explanations until explanatory exhaustion. Aristotle and Thomas of Aquinas used the threatened infinite regress of causes of causes to demonstrate the existence of an unmoved mover, but a child recognizes the game could go on forever.

The central conceit of *The Life and Opinions of Tristram Shandy, Gentleman* (Sterne 1767) is that the novel sets out to recount the life of its hero but, to place him in context and to explain the causes of his character, the text wanders with digressions, and digressions on digressions, with a regress of causes and their causes so that, in the end, we learn little about the eponymous Tristram. A seminal event occurs at the moment of Tristram's conception, when his mother asks his father, "Pray, my Dear, have you not forgot to wind up the clock?" This question "scattered and dispersed the animal spirits, whose business it was to have escorted and gone hand in hand with the *HOMUNCULUS*, and conducted him safe to the place destined for his reception." Many oddities of Tristram's character stemmed from this minor, but far from inconsequential, perturbation.

It is not implausible, indeed it is probable, that whatever my father and mother were thinking during the consummatory act before my conception had an influence on their posture and on which of the myriad sperm in my father's ejaculate won the race to the ovum in my mother's oviduct. "Replaying the tape of life" retells a story in every detail because the sequence of causes remains unchanged. But the first time the tape was played, there was no way of knowing, until it happened, which of my father's sperm would fertilize my mother's egg. There is a single causal narrative out of the past but a beyond astronomical proliferation of possibilities into the future. One can explain with much greater confidence than one can predict.

Taking a step further back from my conception, a complex convergence of molecular events determined the location of chiasmata in the spermatocyte that gave rise to my haploid paternal progenitor. If any one of thirty-odd chiasmata had occurred a mere megabase to either side, then the child conceived would not have inherited my particular set of genes, and the same will have been true of the conception of every one of my ancestors. But a molecular explanation of the location of untold chiasmata would comprise only an infinitesimal part of what a complete causal account of my ancestry would entail. My father's father was an ambulance driver at the second battle of Villers-Brettoneux. So, an account of his survival, where so many others died, would need to explain the trajectories of innumerable projectiles and their fragments, and so on to the endlessly disputed causes of the First World War.

The point of this *reductio ad absurdum* is that, while all evolutionary processes are, in principle, reducible to physical causes, no feasible account can be causally complete. Every story needs a place to begin which leaves many things unsaid. So too, all scientific explanations include items that, for present purposes, are accepted without explanation.

Aristotle Redux

> It were infinite for the law to judge the cause of causes, and their impulsions one of another; therefore it contenteth itself with the immediate cause, and judgeth of the acts by that, without looking to any further degree.
>
> —Francis Bacon (1596)

In pre-Classical Greek, *aition* and *aitia* had connotations of responsibility, guilt, blame, and accusation (Frede 1980; Pearson 1952). Aristotle's *aitia* was translated as classical Latin *causa*, a word that could refer to a lawsuit as in *nemo iudex in causa sua* ("No man should be a judge in his own cause"). English *cause* was adopted from medieval Latin around 1300 and retains legal uses as in *probable cause*. A similar association of cause and culpability occurs in Germanic languages. German *Ursache* (cause, reason, or motive) is related to Anglo-Saxon *sake* as in *for the sake of*. "Sake" could refer to a lawsuit, complaint, accusation, or guilt. Thus, concepts of cause appear to have evolved from proto-legal notions of blameworthiness. A cause was something that could be held responsible.

Aristotle recognized four kinds of *aitia*; traditionally translated as material, efficient, formal, and final causes. Bacon (1605) embraced material and efficient causes as the proper domain of physics but banished formal and final causes to the realm of metaphysics. Aristotelian pluralism was supplanted by a monistic concept of causation of which efficient cause was the dynamical aspect and material cause the physical substrate. In the new mechanical philosophy, form lacked independent potency but was "confined and determined by matter." Final causes were disparaged as an encumbrance to the advancement of learning, as "*remoraes* and hindrances to stay and slug the ship from further sailing."

The fundamental incompleteness of all causal stories has coexisted with faith in explanatory reduction because of scientists' confidence that a physical explanation could, in principle, be given of the things that are left unexplained in each particular causal account. For logical consistency, it should be scientifically and philosophically legitimate to invoke things that look like formal or final causes if these could, in principle, be explained by physical and material causes. The original intent of the paper that became this chapter was to defend the use of formal causes (information) and final causes (functions) in evolutionary explanation, but my purposes evolved in the writing. Formal causes will be presented as abstractions of material causes and final causes as efficient ways of talking about efficient causes. Form can be grounded in material cause because the matter of evolved beings possesses intricate fine structure that embodies experience of what has worked in the past. Purpose can be grounded in efficient cause because current means are explained by past ends via the recursive physical process we call natural selection.

Eggs and Chickens

> Let's think of eggs.
> They have no legs.
> Chickens come from eggs
> But they have legs. The plot thickens:
> Eggs come from chickens,
> But have no legs under 'em.
> What a conundrum!
> —Ogden Nash (1936)

Consider a causal chain: A causes B causes C causes D causes E. Prior things cause posterior things. C is an effect of A and B but a cause of D and E. So much is simple. But what happens when things recur? . . . A_{i-1} causes B_{i-1} causes C_{i-1} causes D_{i-1} causes E_{i-1} causes A_i causes B_i causes C_i causes D_i causes E_i causes A_{i+1} causes B_{i+1} causes C_{i+1} causes D_{i+1} causes E_{i+1}, with the recursion continuing into the indefinite past and indefinite future. Tokens of each type occur both before and after tokens of each other type. A token is either cause or effect of another token—it cannot be both—but cause and effect are inextricably entangled once one attempts to generalize and describe lawful relations among types. Types are both causes and effects of each other (and of themselves). A linear causal chain was chosen for simplicity of exposition but similar arguments could be developed for multidimensional causal webs.

"Self-evident" distinctions between cause and effect are far from obvious in recursive processes. As one moves back along a chain of physical causation, one encounters things that resemble things to be explained. Eggs produce chickens and chickens produce eggs. Genes are causes of phenotypes and phenotypes causes of which genes replicate. What sound is input and what sound is output when an amplifier feeds back?

A phenotypic effect (P) may be viewed as both a cause and consequence of a genotypic difference (G) when both are considered as types. A complete causal account of P_i (subscripts indicate tokens) would include many prior occurrences of P plus many prior occurrences of G and would resemble a complete causal account of G_i. If P_{i-1} causes G_i causes P_i causes G_{i+1}, then it is a matter of preference whether P is considered the cause and G the effect or the other way round. A molecular biologist argues from G to P when explaining how gene expression determines phenotype whereas an evolutionary biologist argues from P to G when explaining why a gene has its particular effects. The former mode of explanation is commonly accepted as unproblematic whereas the latter is rejected as teleological and unscientific. But this is no more than a convention of scientific storytelling. Phenotypes are among the efficient causes of genotypes (the central dogma of molecular biology notwithstanding).

Two other points are worth making briefly. First, a recursive non-equilibrium system must be thermodynamically open because a closed system cannot return to an earlier state (the entropy of the closed system increases until thermal equilibrium). Second, evolution requires heritable imperfections of recursion or nothing can change.

Retrorecursion

> Information can exist only as a material pattern, but the same information can be recorded by a variety of patterns in many different kinds of material. A message is always coded in some medium, but the medium is really not the message.
> —George C. Williams (1992)

Most eukaryotic genomes harbor retroelements that replicate their DNA via RNA intermediates or, what amounts to the same thing, replicate their RNA via DNA intermediates. Nothing structural persists in this process. DNA is "copied" into RNA and then RNA is "copied" into DNA at a new location in the genome (Finnegan 2012).

An LTR retrotransposon can serve as a paradigm. In its guise as double-stranded genomic DNA, the retrotransposon is transcribed by host-encoded RNA polymerase from an antisense-strand of DNA into a sense-strand of RNA. The resulting RNA can have two functional fates: it can be processed into a messenger

RNA (mRNA) that is translated by ribosomes into gag and pol proteins; or it can be used as genomic RNA that is packaged with pol and gag proteins as an infective particle. Pol is a remarkable gadget: acting as a reverse transcriptase, pol synthesizes an antisense-strand of DNA complementary to the genomic RNA; acting as an RNAse, pol degrades the RNA template; acting as a DNA polymerase, pol synthesizes a sense-strand of DNA from the antisense-strand; and acting as an integrase, pol inserts the double-stranded DNA into a new site in "host" DNA (Finnegan 2012). A sense-strand of RNA can be used as a template to make proteins (translation) or antisense DNA (transmission), but the same copy cannot perform both functions.

Retrotransposons trace their origins back before the beginning of cellular life, but an active retrotransposon cannot reside long at any one place in the genome. At each location its DNA is inserted, natural selection favors mutations that inactivate and degrade retroelement functions because retrotransposition is costly to organismal fitness. Nevertheless, retrotransposition persists, because reverse-transcribed DNA inserts at new sites faster than mutations degrade source DNA. Mutations that enhance transposition disperse to new sites while mutations that reduce transposition accumulate at old sites. An active element must stay one jump ahead of inactivating mutations. It is a restless wanderer, leaving crumbling genomic footprints at each step along the way (Haig 2012, 2013).

Retrotransposition involves changes in substance and material form. Consider a nine-nucleotide segment of *gag*. 5′–CGCACCCAT–3′ (antisense DNA) is transcribed into 5′–AUGGGUGCG–3′ (RNA), which can be translated as methione–glycine–alanine (peptide) or reverse transcribed as 5′–CGCACCCAT–3′ (antisense DNA). The latter is then used to synthesize 5′–ATGGGTGCG–3′ (sense DNA). Sense and antisense DNA differ, not only in the use of complementary bases, but also because complementary bases occur in reverse order relative to the sugar-phosphate backbone because of antiparallel pairing. Sense DNA and RNA differ in the substitution of thymine (T) for uracil (U) and in the use of deoxyribose rather than ribose in the backbone. RNA and peptide are chemically chalk and cheese.

The above paragraph was written to disconcert readers familiar with the conventions of representing Watson–Crick base pairing because all nucleotide sequences were written in the direction 5′ to 3′ (the direction of synthesis of the individual strand). Sense and antisense sequences are synthesized in opposite directions with the 5′ end of one complementary to the 3′ end of the other (antiparallel pairing). For this reason, one sequence is usually represented in the 5′ to 3′ direction and the other in the 3′ to 5′ direction so that the complementary bases (A with T and G with C) occur at the same relative position in the represented sequences. My aim in violating the representational convention was to emphasize the chemical differences between a sense sequence and its interpretation as an antisense reverse complement.

Many things within cells are made of DNA, RNA, or protein. Many RNAs are transcribed and many proteins translated. What allows us to pick out a retrotransposon as a nameable entity from these other components and activities? What thing can be held responsible? The retrotransposon is distinguished from other cellular components because it possesses distinct criteria for evolutionary success. Sense DNA, antisense DNA, sense RNA, and peptide are linked by complex causal dependence but are structurally unrelated. Each can be considered to *represent* the others as material avatars of an immaterial gene. The "information" that *is* the retrotransposon must repeatedly change substance and location to persist in an unbroken chain of recursive representation. Representation *presents again* in different form, but prior forms are *present again* when forms recur. There is, in principle, a complete causal account that invokes nothing but efficient and material causes, and in which there is recurrence without continuity of any material thing, but one cannot give a meaningful account of a retrotransposon without reference to its *telos* and *eidos*. The forms are shadows of shadows.[1]

Formal Causes and Information

> If it be true that the essence of life is the accumulation of experience through the generations, then one may perhaps suspect that the key problem of biology, from the physicist's point of view, is how living matter manages to record and perpetuate its experiences.
> —Max Delbrück (1949)

Medieval Latin *informatio* referred to molding or giving form to matter (Capurro and Hjørland 2003): a potter informed the clay. Anglo-Norman *informacione* (13th cent.), however, was a criminal investigation by legal officers. Metaphors of information abound in modern biology. Not everyone who uses them is a fool. There must be meaning behind the metaphors, but precisely what that meaning is has been difficult to pin down. Max Delbrück wrote that "unmoved mover" "perfectly describes DNA; it acts, creates form and development, and is not changed in the process" (1971, 55). Biological information, whatever that may be, performs an explanatory role similar to Aristotle's *eidos* (Grene 1972).

1. This is my footnote to Plato.

An evolutionary distinction between information and objects in which information resides has often been made. It appears in contrasts between replicators and vehicles (Dawkins 1976), information and its avatars (Gliddon and Gouyon 1989), codical and material domains (Williams 1992), and my distinction between informational and material genes. In my formulation, material genes are physical objects, but informational genes are the abstract sequences of which material genes are temporary vehicles. I have previously identified material genes with gene tokens and informational genes with gene types, but the latter is not quite right if "type" is interpreted as a material kind. Sense DNA, antisense DNA, RNA, and protein all represent an informational gene but are not molecules of one kind. Continuity resides in the recursive representation of immortal pattern by ephemeral avatars.

Shannon information quantifies the reduction of uncertainty for a receiver observing a message relative to other messages it could have been. The larger the set of possible messages, the greater the reduction in uncertainty. Perhaps a better formulation would be to say that information measures the reduction of uncertainty of an *interpreter* observing one thing rather than other things it could have been. The interpreter uses the observation to select an interpretation from a set that matches possible interpretations to possible observations. In this formulation, a message (or text) corresponds to the special case of information sent with intent, but an interpreter can also observe the environment or things intended to be hidden.

A human genome contains 3.2 gigabases (Gb) with up to two bits of information per base (a choice from four alternatives). Therefore, a human genome contains 6.4 gigabits of information relative to the set of all possible 3.2 Gb strings. This is the reduction in uncertainty provided by a particular sequence for an interpreter who had no prior knowledge other than the length of the sequence. Every 3.2 Gb string contains the same information but most strings cannot be meaningfully interpreted (Moffatt 2011; Winnie 2000). Only an infinitesimal subset of the library of all 3.2 Gb sequences contains genomes that have ever existed (Dennett 1995). Other measures of Shannon information might compare the sequence to the set of all extant human genomes or to the set of all past genomes. The amount of Shannon information depends on the background knowledge of the receiver.

Information and meaning are distinct. A DNA sequence contains information that acquires meaning when the sequence is interrogated for answers to particular questions. One might use it to determine the amino acid sequence of an otherwise unknown protein or to search for the cause of genetic disease in a patient. Genomes contain clues about evolutionary history if we can only read the hints. If an individual carries the Benin sickle-cell *S* haplotype, then we can infer that he or she had recent ancestors who lived in West Africa and survived malaria. Other inferences can be made by comparing sequences. We compare DNA documents to reconstruct phylogenetic trees, to date times of divergence, to infer ancestral population size, or to locate regions of positive selection.

Information has meaning *for an interpreter* when it is used *to achieve an end*. The proximate end of the interpretative process is an *interpretation* of the information. Interpretation of one thing as another differs from simple change of one thing into another, because interpretation has an intended end. An interpretation is intended for use, but an uninterpreted change simply occurs. This account of meaning can be viewed as parallel to C. S. Peirce's (1877) account of belief. His trinity of belief, desire, and action—"our beliefs guide our desires and shape our actions" (5)—can be loosely translated as my triad of meaning, end, and interpretation. For Peirce, beliefs were habits of mind that guided action: "Belief does not make us act at once, but puts us into such a condition that we shall behave in some certain way, when the occasion arises" (6). In other words, beliefs were latent information whose meaning was expressed in conditional action to achieve a motivated end.

Meaning resides in the interpretation, not in the information, because the same information can mean different things to different interpreters. A sender may intend a particular interpretation and have constructed a message accordingly, but the recipient determines how the message is interpreted. Subsequent interpreters of a message may obtain more, or less, information than was intended by the sender.

Meaning is extracted from a DNA sequence, represented in the output of an automatic sequencer, when a technician reads T rather than A and infers that a fetus will express hemoglobin S. The technician's end is clinical diagnosis. Meaning is extracted from the same DNA sequence, represented as an RNA message, when a ribosome incorporates valine rather than glutamate into a β-globin chain. The ribosome's end is protein synthesis. Selectively neutral single-nucleotide polymorphisms have meaning for a geneticist who uses them to isolate a disease-causing gene but no meaning for the organisms from which they come. No meaning is extracted when DNA is eaten by a bacterium. The use of something as an object (throwing a stone), rather than as a representation (reading a stone tablet), does not count as use of information.

A pause is in order. A thing contains *information* when it differs from something else it could have been. Two things contain mutual information if an observer can learn something about one by observing the other. This is a symmetric relation. An effect *represents* its cause when observation of the effect allows inference of the cause. This is an asymmetric relation: X_i represents Y_i to the extent that Y_i is causally responsible for their mutual information. A thing has *meaning* for an interpreter when its "difference from something else" is used by the interpreter to achieve an end. An *interpretation* is a representation of the information used by the interpreter.

An interpretation can be a text interpreted by another interpreter. Interpretation is recursive when interpretations return to prior forms. X and Y, considered as types, reciprocally represent each other if the token X_i represents Y_i represents X_{i-1} represents Y_{i-1}. Replication is reliable, high-fidelity recursion of interpretation. (The game of "Telephone" shows what happens when representation is unreliable.) The text of a replicator is an interpretation of itself.

Living things are replete with reliable reciprocal representation. Each strand of the double helix represents the other. A messenger RNA (mRNA) represents the DNA from which it is transcribed, and the DNA represents the mRNA. A protein represents the mRNA from which it is translated, and the mRNA represents the protein. DNA represents protein, and protein represents DNA. Extended phenotypes represent genotypes, and genotypes represent extended phenotypes (Dawkins 1982; Laland et al. 2013a). All represent what has worked in past environments. Natural selection creates complex causal dependence between past environments and processes within cells.

Life is made meaningful by a multitude of mindless interpreters reinterpreting the molecular metaphors of other mindless interpreters. RNA polymerases transcribe DNA as RNA. tRNAs interpret codons as places to deposit amino acids. Ribosomes translate RNA prose into protein poetry. Higher-level interpreters depend on the activity of myriads of lower-level interpreters. Islet cells integrate blood glucose and other inputs to regulate insulin. Fat cells, muscle cells, and liver cells interpret insulin for diverse ends. Neurons respond to signals from muscles and muscles to signals from neurons. Brains comprehend social relations. You read this sentence. Organisms are self-constructed interpreters of genetic texts in environmental context.

The environment chooses phenotypes and thereby chooses genes that represent its choices and embody information *about* the environment's criteria of choice. Observation of these choices would reduce the uncertainty of an omniscient observer about which genes will be transmitted to future generations. The choices of the environment are unintended, but actions that are repeated because of their effects are thereby intended. The choices of the environment are not themselves messages, but genes that represent these choices are copied and passed on as messages from one generation to the next (Bergstrom and Rosvall 2011). Organisms and their lower-level parts are senders and interpreters of these texts.

Difference Demystified

> A difference is a very peculiar and obscure concept. It is certainly not a thing or an event.
> —Gregory Bateson (1972)

A soldier fires at Marius but Éponine blocks the shot with her body, saving Marius's life. The soldier's choice, the difference between firing or not firing, makes no difference as to whether Marius survives but does make a difference as to whether Éponine

survives. Éponine's choice, the difference between lunging forward or holding back, makes the difference between Marius's
death or survival. The soldier's shot is responsible for Éponine's
death and Éponine's death is responsible for Marius's survival,
but the soldier's shot is not responsible for Marius's survival.
Responsibility is not transitive.

Things or events do not make a difference; differences between
things or events make a difference. One cannot decide whether
something is responsible for an outcome without answering the
question, compared to what? A choice is an act that *could* have
been otherwise and *may* make a difference.

A physician gives morphine to a patient dying of cancer. The
difference between a fatal and nonfatal dose does not make a difference between the patient dying or not dying, but does make
a difference between the patient dying a painful or nonpainful
death. If I tell you the dose of morphine I do not provide any
information about whether the patient lives or dies but provide
information about the nature of the death. If the patient does
not die from an overdose, then the patient dies from cancer. In
the philosophical literature, this is known as causal preemption
(Hitchcock 2007).

There is a close connection between concepts of information
and causation. Gregory Bateson (1972) defined the unit of information as a "*difference which makes a difference*," but his phrase
could also be used as a definition of causation with the first difference as cause and the second as effect. (William Bateson, who
coined the word "genetics," named his third son after Gregor
Mendel.) In the words of Ronald Fisher, "To the common sense
of mankind it is the property of a cause, *qua* cause, that it might
have been different and have had different effects" (1934, 106).
Observation of either difference contains information about the

other. This information is potentially *about* the relation between cause and effect, but *use* of the information requires an interpreter that has either been designed or evolved for that end.

Consider again the nine-nucleotide segment of *gag* antisense DNA. When 5'–CGCACCCAT–3' is transcribed as 5'–AUGGGUGCG–3' by an RNA polymerase, every DNA nucleotide makes a difference in the resulting RNA. RNA polymerases are instructed by DNA sequences in which every nucleotide conveys actionable information: A means "choose U," C means "choose G," G means "choose C," and T means "choose A." Once transcription is initiated, and until it terminates, RNA polymerase interprets every A, C, G, or T as U, G, C, or A regardless of the context of surrounding nucleotides. Each and every change in the DNA nucleotide sequence would cause a change in the RNA sequence (given a properly functioning RNA polymerase).

Ribosomes translate 5'–AUGGGUGCG–3' as methione–glycine–alanine. They are more sophisticated interpreters than RNA polymerases because the meaning of bases for ribosomes is determined by context. The AUG triplet communicates crucial information. It is the symbol "start here with methionine" that initiates most polypeptides and sets the reading frame for translation of the rest of the message in triplets. AUG in the body of an mRNA (when in the correct reading frame) simply means "choose methionine." The two meanings are distinguished by context.

G appears five times in the sequence of nine bases. The G in AUG is essential for meaning "choose methionine" because any other base in that position would result in a different amino acid added to the polypeptide. The two Gs in GGU taken together mean "choose glycine" because GGC, GGA and GGG are also interpreted as glycine by the ribosome. The first G in GCG

means "choose alanine" in the context of C in the second position, because any other base in the first position would be interpreted as a different amino acid; the G in the third position does
not make a difference and could be replaced by any other base
and still be interpreted as alanine; but a deletion of the third
base (a difference between nobase and somebase) would shift
the reading frame and change the interpretation of the rest of
the message.

RNA polymerases and ribosomes choose from ensembles.
When an RNA polymerase transcribes G, it picks out a C from a
cytoplasmic mixture of U, C, A, and G. When a ribosome translates AUG, it selects a tRNA charged with methionine from a
mixture of tRNAs charged with all twenty amino acids. Methionine is the *bon mot* the ribosome seeks to capture the meaning
of AUG. AUG is present in this position in the RNA message
because it has competed, and will compete, with alternatives
such as ACG or UUG that have different denotations for the
ribosome and different connotations for the organism. Natural
selection among variant texts chooses those that are useful and
discards the dross. By this means, the macrolevel of ecology and
social interactions informs the microlevel of molecules.

Some changes to an RNA message change the amino acid
added to the growing polypeptide—these are differences that
make a differance in the translated protein—whereas other
changes are synonymous and make no difference in translation.
The choice of a particular amino acid at a particular location in
a protein may have no effect on protein function, in which case
different codons are meaningful for the ribosome but meaningless for the organism. For such a "neutral" substitution, the difference in the mRNA, and in the DNA from which the message
was transcribed, *causes* a difference in the protein but *does not*

cause a difference in fitness. The choice of amino acid by the ribosome is purposive, but the choice of nature is random.

A choice is a difference that makes a difference. It is a branch point at which a traveler could have taken another path but, once a path is chosen, the chosen path informs an observer of the traveler's choice. Information about what befalls on a path would be useful in making a choice if the traveler ever came that way again. If travelers copy their choices for later reference, and death awaits on one path but safety on another, then choices of the wrong path never return to the fork in the road, but choices of the right path return to make the same "wise" choice again. In a perilous maze, the records of surviving travelers provide a safe guide for finding a way.

Choices are degrees of freedom. The meanings of information are the choices it guides. Information is useful if, and only if, it changes the future for the better. By tortuous paths, we have come to view choice as synonymous with cause and information as a potential guide of choice. Given a textual record of recurring choices, Darwin's demon (Pittendrigh 1961) culls the bad choices and retains the good. Well-informed choice is purposive difference-making.

Final Causes and Functions

> It follows that there are several causes of the same thing. . . . And things can be causes of one another, e.g. exercise of good condition, and the latter of exercise; not, however, in the same way, but the one as end and the other as source of movement.
>
> —Aristotle, *Metaphysics*

Teleological language in biology appears in a heterogeneous class of explanations united by the loose property that a thing's existence is explained by an effect that the thing makes possible. A beaver grows sharp incisors to cut down trees to build a lodge to provide shelter from the storm. Dental development has the *goal* of sharp incisors with the *function* of cutting down trees *for the sake of* building a lodge for the *purpose* of shelter, all *for the good* of the beaver. "In order to gain access to buried stretches of DNA inside nucleosomes, a chromatin remodeling ATPase is required to unwrap the nucleosomal DNA" (Mellor 2005, 147) is no less teleological than "the hairs about the eye-lids are for the safeguard of the sight" (Bacon 1605/1885, 120).

A final cause explains something by its effects. The thing exists for the sake of an end. In the absence of conscious intent, such explanations have been rejected because *explanandum* precedes *explanans*. However, this argument loses force for products of natural selection because $ends_i$ can be causes of $means_{i+1}$ without backward causation. A thing exists today because similar things in the past had effects that enhanced survival and reproduction. The thing expresses similar effects in the present because its effects are heritable. Therefore the thing considered as a type exists because of its effects.

Ends can be means to other ends. Ayala (1970) distinguished proximate ends, the functions or end-states a feature serves, from the ultimate goal of reproductive success. Most biological research addresses the end-directedness of adaptations to achieve proximate ends without explicit reference to ultimate goals. The proximate ends of the mindless interpreters described in previous sections are interpretations of information from the environment or sent as genetic texts. The purposeful behavior of these interpreters can be explained as the outcome of selective processes that incorporated information about what worked in past environments into the fine structure of information-carrying molecules.

Selection means choosing from a set of alternatives. If there is no alternative, there can be no choice. In Darwin's metaphor of *natural* selection, the environment "chooses" via differential survival and reproduction. In my formalism of this process for genetic replicators, the environment chooses among effects of genes and thereby chooses among genes. An effect is a difference a gene makes relative to some alternative. It is not a property of an individual gene but rather is a relation between alternatives. The selected gene is a difference that made a difference. In this formalism, phenotype (synonymous with a gene's effects) is defined as all things that differ between the alternatives, whereas environment is defined as all things shared by the alternatives. By these definitions, what is a phenotype in one comparison may be environment in a different comparison. Natural selection will tend to convert phenotype into environment because environment is that for which there is no reasonable alternative. Deleterious mutations are unreasonable choices that are eliminated by "negative selection." They are difference-making alternatives that are eliminated soon after they occur.

Choices of the environment reduce uncertainty about which genes will leave descendants. The selected genes thereby convey information about these choices to ribosomes and other mindless interpreters in subsequent generations. If the choices of the environment are nonrandom, then the genes embody usable information *about* the environment's criteria of choice and guide effective choices of organisms.

A gene is "responsible" for its effects. Changes of allele frequency extract average additive effects on fitness from a matrix of nonadditive interactions (Fisher 1941). Whatever effects of an allele contribute to a positive average effect on fitness can be considered the final causes of the allele's persistence. A gene's

function can be defined as those of its effects that have contributed positively to its spread and present frequency. All other
effects, negative or neutral, are side effects without function. If
an effect contributes to a gene's success—by any route, no matter how devious—then the gene exists for the sake of that end
and the end exists *for the good of the gene* (Haig 2012; Haig and
Trivers 1995).

In the struggle for existence in a world of finite resources, one
variant's success comes at the expense of alternatives. The causes
of death of individuals without an allele contribute to an allele's
success, just as much as the causes of survival of individuals with
the allele. The less-appealing traits of the suitors rejected by my
mother in favor of my father comprise part of a complete causal
account of how I happen to be writing this essay.

An allele must make a difference in many lives if it is to spread
by natural selection, from a single copy arising by mutation in
a germ cell to fixation in a population of many individuals. No
one event can be singled out as the cause of adaptation, but patterns of events, distributed through space and time, result in
adaptive change. Natural selection is not an efficient cause but a
statistical summary of many efficient causes.

One must consider not only allelic substitutions (positive
selection) but also failures of substitution (negative selection).
All adaptations will degrade over time unless mutations that
impair the evolved function are weeded out. Each new mutation
creates an allelic difference that is subject to selection on the
basis of its average effect on fitness. If the mutation is eliminated
by a choice of nature, then the difference of phenotypic effect
exists for the good of the allele chosen. Many phenotypically
interchangeable but genetically distinct loss-of-function mutations can be grouped together into a single allelic difference. In

this way, a genetic function, determined by interactions between multiple sites within a coding sequence, can be considered *for the good* of the evolutionary gene.

Consider the substitution of thymine for adenine in the middle base of the sixth codon of the human β-*globin* gene. This difference causes a replacement of glutamate by valine at the sixth amino acid position of the β-globin polypeptide. The resulting protein, hemoglobin S, is responsible for sickle-cell disease when homozygous and resistance to malaria when heterozygous. The alternative allele with valine at position 6 is known as hemoglobin A. With respect to the allelic difference between *A* and *S*, the function of *S* is containment of malarial infection in a genotypic environment that includes an *A* allele. A deleterious side effect of *S* is life-threatening anemia in a genotypic environment that includes another *S* allele (Haig 2012).

The prior paragraph deliberately confuses gene and protein. Proteins and genes often share the same name (mutual metonymy). Sometimes a gene is named for its protein and sometimes a protein for its gene. In speech, a gene name often collectively denotes gene, mRNA, and protein as avatars of recursive form.

The sickle-cell mutation has been presented as an exemplar of a "selfish nucleotide" and used to dispute the identification of "evolutionary genes" with DNA (Griffiths and Neumann-Held 1999). The *reductio ad absurdum* fails because evolutionary genes have been defined as stretches of DNA rarely disrupted by recombination (Dawkins 1976; Williams 1966) and sufficiently short to maintain linkage disequilibrium (Haig 2012). Nonrandom associations of variable nucleotides, some of which may be functional, extend for hundreds of kilobases to either side of the "selfish thymine" (Hanchard et al. 2007). As recombination between sites lessens, and as the strength of epistatic selection increases, a point is reached at which different sites can no longer be considered as belonging to different evolutionary genes (Neher, Kessinger, and Shraiman 2013). For sites sufficiently close together, nonadditive interactions on the axis of expression contribute to an additive effect on the axis of transmission (Haig 2011a; Neher and Shraiman 2009).

Any complex organismal adaptation will involve many allelic substitutions at multiple loci. For ancient adaptations, most substitutions will have occurred in the deep past, in organisms and environments very different from those of the present. In the process, some genes may have been transformed beyond recognition. While each substitution could be considered for the good of that gene at that time, the adaptation serves proximate ends today. For what entity are these ends a good? A standard answer is that complex adaptations are *for the good of the organism*. A gene-selectionist could counter that a complex adaptation is for the good of each and every gene whose loss of function by mutation results in loss of the adaptation (Haig 2012).

Darwin's Demon

> The literature written by [Darwin's] Demon is no more deducible from a complete command of the nucleotide language, let alone physical law, than the works of Shakespeare or Alfred North Whitehead are deducible from a complete command of the English language.
> —Colin Pittendrigh (1993)

Chickens can unscramble eggs by eating them and laying another (Gregory 1981, 137). James Clerk Maxwell (1831–1879) imagined a demon that performed work by choosing which molecules to allow through a partition, thereby selecting ordered subsets from a disordered ensemble. Selection can extract work from randomnesss.

A rocket is a rigid tube, open at one end, that converts the disordered molecular motion of combustion into coherent motion of the tube. Roughly speaking, the closed end of the tube *selects* molecular momentum orthogonal to its surface and imparts that momentum to the rocket while the open end *discards* momentum in the opposite direction. The rocket engine is the *selective environment* that chooses an ordered subset of moving particles from a disordered set as the entropy of the working material increases. A piston selects molecular momentum orthogonal to the one moveable wall of a cylinder and thereby does work while discarding unworkable energy into a heat sink (Atkins 1994, 83). Organisms are elaborate self-assembling engines that acquire or synthesize their own fuel and dump entropic excrement. They are the selective environment by which food is converted to work.

Subset selection is a semantic engine. Consider a set subject to a procedure by which some are "chosen" and others "rejected." Choice is random if membership of the selected subset is determined by criteria independent of intrinsic properties of things chosen (for example, if no attribute has a periodicity of five but every fifth entity is selected). The disjunction of selected and discarded subsets contains no information about the criteria of choice when choice is random, but the disjunction contains information about the criteria of choice when choice discriminates among members of a set on the basis of one or more of their intrinsic properties (a reasoned choice). The selected and discarded subsets are biased samples of the whole. One might say that one is adapted, and the other maladapted, to the selective environment.

Wind winnows wheat from chaff by the criterion of weight to cross-sectional area. A bird picks berries from a bush on the basis of palatability, and the bird's criteria of choice are reflected in differences between eaten and uneaten berries. A man chooses a wife and we can infer something about his preferences by comparing his spouse to others who were available but passed over. His choice is restricted to members of a comparison set constrained by the comparison sets and preferences of possible partners. You can't always get what you want.

Natural selection, it has been said, differs from subset selection because "offspring are not subsets of parents but new entities" (Price 1995, 390). But the genes of the next generation are a subset of the genes of the last. Therefore, natural selection can also be inscribed under the rubric of subset selection if focus shifts from vehicles to replicators, from interpretations to texts. Natural subset selection is indirect. The environment selects a subset of phenotypes to be parents and thereby selects a subset of genes to be transmitted.

Selection from a selected subset retains information from past choices, imperfectly. Retention is imperfect because information is dissipated by random culling, by random mutation of past reasoned choices, and by changes in criteria of choice. In the absence of replication, recursive selection reduces the size of the comparison set at each round of choice. Replication creates redundancy and thus increases the probability that information from past choices will be retained despite dissipative forces.

Mutations are random guesses in the neighborhood of previous choices. Mutation degrades semantic information about past choices but adds entropy for future reasoned choice. For the right balance of mutation and selection, *recursive selection of mutable replicators* results in accretion of semantic information and refinement of fit to criteria of choice.

Mendel's Demon

Why all this silly rigmarole of sex? Why this gavotte of chromosomes? Why all these useless males, this striving and wasteful bloodshed?
—William D. Hamilton (1975)

Clonal reproduction replicates entire genotypes that are judged repeatedly in the court of environmental opinion. Each asexual genotype is a single "evolutionary gene" responsible for its own average effects after repeated retesting. The difference between genotypes that differ at a single site can be attributed to that site, but responsibility cannot be attributed to individual sites when genotypes differ at multiple sites. Segments of particular value must share credit with segments that do not pull their weight and are hidden from blame. All must share in communal praise and collective guilt.

Sexual genotypes, by contrast, are ephemeral. Judgment of each individual genotype is unique and unrepeated, but smaller segments are tested repeatedly against different backgrounds and can be held responsible for their average effects. Sexual genotypes are pastiche, cobbled together from parts of two parental genomes, four grandparental genomes, eight great-grandparental genomes (you get the idea), in a process that mindlessly breaks up effective combinations for the chance of something better. Every one of these genomes has been tested by the environment and passed. The sexual disassembly and reassembly of genotypes allows attribution of responsibility to parts.

Mendel's demon (Ridley 2000) is a randomizing agent that shuffles the genetic deck and deals out fresh hands in each round. It can be a mischievous imp that impedes the work of Darwin's demon by breaking up favorable combinations, or a helpful sprite that rescues parts of promise from bad company. As the genome is diced into smaller pieces, the range of effects for which each nonrecombining segment can be held responsible diminishes (Godfrey-Smith 2009, 145; Okasha 2012), but each segment is more readily held responsible for its causal effects. Darwin's and Mendel's demons, working together, create teams of champions rather than champion teams.

Peirce's Demon

> Experiment . . . is an uncommunicative informant. It never expiates: it only answers "yes" or "no." . . . It is the student of natural history to whom nature opens the treasury of her confidence, while she treats the cross examining experimentalist with the reserve he merits.
> —C. S. Peirce (1905)

C. S. Peirce (1905) compared an experimental scientist with men whose education had largely been learned from books: "He and they are as oil and water, and though they be shaken up together, it is remarkable how quickly they will go their several mental ways, without having gained more than a faint flavor from the association" (161). His vivid use of metaphor belied his admonition "that no study can become scientific . . . until it provides itself with a suitable technical nomenclature, whose every term has a single definite meaning universally accepted among students of the subject, and whose vocables have no such sweetness or charms as might tempt loose writers to abuse them" (163–164). He contrasted the poverty of the experimentalist's "meagre jews-harp of experiment" to the richness of the naturalist's "glorious organ of observation" (175). Despite such a seemingly invidious comparison, the rational purport of belief was to be found solely in answers to repeated experiments and their consequences for future conduct: "If one can define accurately all the conceivable experimental phenomena which the affirmation or denial of a concept could imply, one will have therein a complete definition of the concept, and *there is absolutely nothing more in it*" (162). Right conduct is choice guided by experience.

Experiments are choices offered to nature for the resolution of doubt. They provide terse, inarticulate answers to narrowly defined questions. These answers are informative when they reduce the experimentalist's uncertainty about the state of the world. The beliefs they engender have meaning when used to guide conduct. By this means, "thought, controlled by a rational experimental logic, tends to the fixation of certain opinions" that are not arbitrary but predetermined by nature (Peirce 1905, 177).

The experimental method (Peirce's demon) and natural selection (Darwin's demon) are resolvers of difference in which choices of nature inform adaptive behavior via the accumulation of useful information. Practice perfects performance by trial and choice. A controlled experiment varies one thing while holding other things constant (*ceteris paribus*) to determine the differences for which that thing can be held responsible. But experiments must be replicated to average out residual, uncontrolled variation. Sexual recombination achieves a similar statistical control by repeated retesting of allelic differences on different genetic backgrounds. The average effects of allelic differences reduce the complexity of biological interactions to simple binary choices. The success of the experimental method and of sexual organisms suggests that short-sighted choice among recombinable units often outperforms reasoned judgment of integrated wholes.

The histories of causal and legal concepts are closely intertwined. The function of a trial is to determine whether a defendant is responsible for a crime. Many circumstances and opinions are weighed in the balance but the judgment is binary, guilty or not guilty. The earliest known meanings of *try* are to sift or pick out, to separate one thing from another, especially the good from the bad, and to choose or select. A *trial* was the determination of a difference, between guilt or innocence, by tribunal, battle, or ordeal. Natural selection is a recursive process of trial and judgment by which good causes are rewarded and relative truths learnt.

Gene-Selectionism and Developmental Systems Theory

> A thing exists as a natural end *if it is cause and effect of itself.*
> —Immanuel Kant (1790/2000)

Phenotype interprets genotype in environmental context. Why should genes be singled out as possessors of purposes and as self-interested beneficiaries of adaptation? Genes belong among the material causes of development, and gene expression among the efficient causes of development, but ontogeny proceeds via complex interactions between genes and environment. From the perspective of developmental systems theory, the causal matrix recreates itself, recursively, without a privileged role for genes (Oyama 2000).

Genes interact with each other and the environment to create phenotypes that causally influence which individuals leave descendants. But, when the environment chooses one allele rather than another, the choice is based on the average effect of a difference (Fisher 1941). In Lewontin's (2000) terminology, the allelic effects are causes of difference but the interactions are causes of state. The prosaic selection of differences is the unwitting author of poetic changes of state.

Gene-selectionism is concerned with how information gets into the genome via natural selection and what can be held responsible for the appearance of purpose in nature. By contrast, developmental systems theory is concerned with understanding ontogenetic mechanisms. One might say that gene-selectionism addresses the writing, and developmental systems theory the reading, of a text. From this perspective, the two frameworks are complementary. Any text of lasting value is read, and judged, repeatedly as it is revised.

Two domains of explanation are in play that have been characterized as a vertical axis of transmission and a horizontal axis of development (Bergstrom and Rosvall 2011). One concerns the inheritance of genetic *information* between generations and the other the expression of genetic *material* within generations. Teleological concepts appear in both domains. On the axis of transmission, final causes appear as adaptations that serve the ultimate end of fitness. On the axis of expression, final causes appear as end-states of developmental processes and as the proximate ends of goal-directed behaviors. Explanations in the two domains have different flavors because mapping from gene copy to gene copy in the course of transmission is straightforward whereas mapping from genotype to phenotype in the course of development is devilishly difficult.

The conceptual separation of axes of transmission and development is related to Shea's (2007) separation of phylogenetic and ontogenetic explanations; to Ayala's (1970) distinction between ultimate goals and proximate ends; to Weismann's (1890) separation of germ plasm and cytoplasm; to the difference between DNA replication and RNA transcription; to the divide between text and interpretation; and the contrast between mention and use of a lexical item. Kant (1790/2000, 243) can be interpreted as making a related distinction when he describes the twofold sense in which a tree is both cause and effect of itself. A tree generates itself both as a species/genus (transmission) and as an individual (development).

Whether conceptual separation of developmental from evolutionary questions is productive or counterproductive is a subject of present polemics. Some maintain the distinction is indispensable (Griffiths 2013), whereas others see it as an impediment to understanding (Laland et al. 2013a). Most of those who support the distinction are comfortable with invoking functions as causes, whereas many of those who want to do away with it are explicit that "functions are not causes . . . the outcome of a behavior cannot determine its occurrence" (Laland et al. 2013b).

Our penchant for dichotomies, distinctions, and oppositions reflects the power of reducing complex questions to binary choices. Many arguments within the philosophy of biology, and between the sciences and humanities, reflect a tension between the reductive simplicity of average effects and the richness of interaction; between the meager trump of attributing credit to parts and the glorious Wurlitzer of integration of wholes. But we have more than two options; one can play a duet. (Perhaps I should explain one of my more obscure word choices. A trump is an old name for a jew's harp. My use of trump alluded to Peirce's [1905] juxtaposition of the "meagre jews-harp of experiment" and the "glorious organ of observation." I had no foresight of the outcome of the 2016 presidential election.)

Genomes as Texts

> Are God and Nature then at strife,
> That Nature lends such evil dreams?
> So careful of the type she seems,
> So careless of the single life.
>
> —Alfred, Lord Tennyson (1849)

Genomes resemble historical documents (Pittendrigh 1993; Williams 1992, 6). Thymine rather than adenine, or valine rather than glutamate, has no meaning out of context; but a nucleotide sequence of β-*globin* with thymine at position 17, or an amino acid sequence of β-globin with valine at position 6, both have meaning in context, although neither says anything explicit about malaria. Genomes are allusive archives of choice, with unstated meanings without explicit expression or discrete location. They are palimpsests on which new text is written over partially erased older text (Haig and Henikoff 2004). Not all of the text is readable. It contains gobbledygook and epigenetic annotations that proscribe what should not be read. Genomic censors strive to shut down the clandestine presses of retrotransposons.

Where does meaning reside in a text? This chapter evolved via incremental rewording and extensive rewriting. There was a struggle for existence among ideas for space on the page. There is a lot more *I could have* said. My meaning resides in the difference between what is said and unsaid. Often a change in one part necessitated changes in other parts to maintain consistency. This chapter self-consciously reflects back upon itself with repetition, recurrence, reciprocal reference, and allusive alliteration. Part of its meta-meaning is that many meanings are distributed throughout the text, never fully explicit, to reflect and suggest the organization of meanings within the genome. There is no meaning in a letter, a little in a word, a bit more in a sentence, but much of the intended meaning is implicit, to be understood from the synergistic whole rather than the additive parts. And yet, the text was written letter by letter and word by word by additive increments. On the axis of reading, new meanings can be found, but on the axis of transmission it is only that which is written that counts.

Meaning resides in the interpretation. There are meanings I intend you to find and meanings you find. I wrote to persuade. But you may use my prose to persuade others that I am mistaken. You interpret my text as you will. Imprecision of language allows charity of interpretation and slaying of straw men. Falsehood can arise from misinformation by an author or from misinterpretation by a reader.

The question of what genes mean, if what genes do depends on interactions with other genes in environmental context, resembles the question of what words mean when all definitions are expressed in other words in semantic context. Modern philosophers confront the "indeterminacy of translation" when attempting to understand what *aition* meant to Aristotle and "indeterminacy of interpretation" when attempting to understand, or deliberately misunderstand, each other's arguments.

Modern biologists confront similar indeterminacy in the semantic content of genetic material. Critics of "information talk" in
biology often demand a more rigorous justification of meaning
in DNA than they could provide for meaning in language.

An idea is the semantic equivalent of a nonrecombining segment of DNA. It is a chunk of meaningful stuff that is transmitted as a parcel. It is a semantic difference that makes a difference.
Ideas and "pithy quotations" are readily reusable because they
are meaningful when taken out of context. Science proceeds
via recombination of ideas, whereas great works of literature are
clonally replicated and interpreted as wholes. In the scientific
literature, "smallest publishable units" have replaced magisterial
tomes in part because shorter texts are more likely to be used and
cited. Working biologists mostly read *On the Origin of Species* for
virtue or pleasure, because the good bits have been reused again
and again, in new associations, in a sesquicentury of scientific
endeavor.

There are parallels between the ascription of effects to genes
and the assignment of credit to authors. Scientists cite each
other more than philosophers, and novelists hardly at all. Cita-
tions not only provide pointers to additional information but
also ascribe credit. All new insights originate in the context of
many acknowledged and unacknowledged precursors, but credit
is easier to attribute, or harder to deny, for portable ideas than
for rearrangements in the tangled web of meanings. *Tristram
Shandy* contains philosophical insights but is rarely cited by philosophers because discrete ideas are difficult to disentangle from
its interwoven fabric.

Scientists care about citation because they want their name
to ride the coattails of successful ideas to feedback for their
good. But to be worthy of credit one must be unambiguous.

Otherwise one could claim credit for interpretations that prove prescient but shift blame for interpretations that fail. A scientist is expected to commit to one interpretation, but a novelist often leaves a choice for the reader. Indeterminacy of interpretation is a designed feature of novels but a flaw in experimental notebooks and scientific papers.

Teleodynamics

> In an indeterministic world natural causation has a creative element, and science is interested in locating the original causes of effects of special interest, and not merely in pushing a chain of causation backwards *ad infinitum*.
>
> —Ronald Fisher (1934)

Consider the fates of zygotes, scions of countless spermatic races to ova. Their lives unfold via interactions among genes, and between genes and environment. Many fall by the wayside, by chance or necessity, and those that reach maturity produce progeny, some a hundredfold, some sixtyfold, some thirtyfold. Sometimes an allelic difference causes one to leave more issue than another. And, lo and behold, the genes of the progeny, and of the progeny's progeny, even unto the third and fourth generation, are a biased sample of the genes of their progenitors. The tale is repeated, with minor variations and mutations, time without end, and verily there is something new under the sun.

This evolutionary parable could be elaborated endlessly with causal explanations of ever-finer detail and ever-deeper regression into the past. There is a causal story behind each and every mutation, each and every chiasma, each and every choice of a mating partner, each and every union of gametes, each and every catastrophe that did not happen. But this story is untellable because of incomplete information, chaotic dynamics, and computational complexity. And if it could be told, the story would be incomprehensible. One must simplify to tell a tale, giving greater salience to some items and leaving loose ends.

A pedant could argue that pressure is not an efficient cause and should be expunged from physical explanations—only individual molecular impacts are truly causal—but his argument would be dismissed as obfuscation. For questions at the appropriate scale, pressure provides a perfectly adequate explanation, indeed one that is superior to the unattainable account that describes each and every molecular collision. Darwinian final causes are similarly grounded in efficient causes and are perfectly adequate, indeed indispensable, for certain kinds of biological explanation. A "selection pressure" summarizes many reproductive outcomes just as the pressure of a gas summarizes many molecular motions. Darwinism, like thermodynamics, is a statistical theory that does not keep track of every detail (Fisher 1934; Peirce 1877).

Much recent semantic work has been undertaken on concepts of Darwinian information (Adami 2002; Adami, Ofria, and Collier 2000; Colgate and Ziock 2011; Frank 2009, 2012). The various expositions exhibit phenotypic resemblance, both from shared ancestry and convergence in a common selective environment, although conceptual differences remain. Rather than choose among the differences, I will synthesize a subset of select conclusions. Semantic information comes from the environment via subset selection and refers to that environment. It is functional, looking backward to what has worked in the past and forward as a prediction of what will work in the future.

Replication is essential for the indefinite persistence of information in the face of dissipative entropic forces.

Back to the Future

> The word "cause" is so inextricably bound up with misleading associations as to make its complete extrusion from the philosophical vocabulary desirable.
> —Bertrand Russell (1913)

My intent in partial rehabilitation of formal and final causes is not to argue that the four causes provide the best causal taxonomy for current ends, but to recognize that Aristotle's classification was found useful for more than a millennium and must surely have approximated significant categories of understanding. Moreover, if formal and final causes do not exist in their "bad" metaphysical senses, then the terms and the concepts are available for use in their "good" post-Darwinian senses of inherited information and adaptive function.

This chapter concerns the seduction of narrative, the magic of metaphor, and the rhythm of recursion (Hofstadter 1979). Meaning is expressed through metaphor by representing one thing by another. Recursive representation allows *eidos* and *telos* to be grounded in *hyle* and *kinesis*. Choice captures information. The environment, personified as natural selection, chooses ends and thereby chooses means with meanings, because the ends of the past are the means of the present. Meaning requires an interpreter and an end. Darwin's demon supplies both. My text returns repeatedly to etymologies and histories of ideas because *logos* and *eidos* evolve by paths parallel to genes, providing fruitful metaphors and philosophical perspective.

Natural selection is both a metaphor and a metaphorical process of recursive representation. It is a meaningless, purposeless, physical algorithm that produces things for which meaning and purpose are useful explanatory concepts (Dennett 1995). Among the products of natural selection are rational agents, with beliefs and desires, pursuing conscious goals, exchanging truthful and deceptive information, who can delight in a meaningful life.

L—d! said my mother, what is all this story about?—A COCK and a BULL said Yorick—And one of the best of its kind I ever heard. (Sterne 1767, finis)

Interlude

> Even the most genuine and pure tradition does not persist because of the inertia of what once existed. It needs to be affirmed, embraced, cultivated. . . . Even where life changes violently, as in ages of revolution, far more of the old is preserved in the supposed transformation of everything than anyone knows, and it combines with the new to create a new value.
>
> —Hans-Georg Gadamer (1992)

An interlude (Latin *inter* "between," *ludus* "play") was a short performance between the acts of a long morality play. The previous chapter addressed the origin of purposeful beings whereas the next chapter will address how purposeful beings make sense of their world. This interlude marks a transition from a focus on differences that make a difference in diachronic (evolutionary) time to differences that make a difference in synchronic (behavioral) time.

Recall Darwin's thought experiment of old springs, wheels, and pulleys that have been employed in a series of different contrivances. What is the function of a pulley? One story might pronounce that the pulley's *real* function is the role it performed in the first contrivance of which it formed a part. Another story might declare that the pulley's functions are the roles it has performed in each of the successive devices in which it has been employed and that its current function is the role in which it is currently employed. Yet another story might aver that the functions of pulleys are the causal role they play in changing the direction of a force independent of the purposes for which any particular pulley is employed. What sayest thou?

Natural selection has two aspects that have been labeled *positive selection* (associated with *origin* of novel function and elimination of the old) and *negative selection* (associated with *maintenance* of existing function and elimination of the new). Under positive selection, old less-adapted gene sequences are replaced by *new* more-adapted sequences. Under negative selection, new less-adapted sequences are continually eliminated, soon after they arise, preserving *old* more-adapted sequences. Bouts of positive selection are invariably accompanied by a background of new deleterious mutations that are eliminated by negative selection. Old sequences now maintained by negative selection were once new variants favored by positive selection. Genetic differences make a difference in fitness under both positive and negative selection. The genetic sequences we see are the survivors of both kinds of selection.

Many differences of opinion in evolutionary theory hinge on whether negative selection is included or excluded as a component of adaptation by natural selection. Adaptationists and structuralists agree that conserved morphological and genomic features are maintained by negative selection. They agree that the gene regulatory networks responsible for developmental constraints are conserved, not because mutations do not occur, but because mutations are not tolerated. The problem is that adaptationists consider these to be functional constraints because they include both positive and negative selection under adaptation by natural selection, whereas structuralists consider them to be structural constraints because they exclude negative selection from adaptation. These semantic differences have been a recipe for mutual misunderstanding. As a consequence, adaptationists *interpret* the maintenance of structure in the face of mutation as an expression of the power of natural selection, whereas structuralists *interpret* the absence of positive selection as evidence of a constraint on what can evolve.

A definitional restriction of adaptation to positive selection is often accompanied by claims that a feature is only an adaptation for the "original" function for which it evolved by positive selection but not for any of the subsequent uses in which it is employed, even though performance of these new roles is maintained by negative selection. Thus, Stephen Jay Gould and Elisabeth Vrba (1982, 6) defined an adaptation "as any feature that promotes fitness and was built by selection for its current role. . . . The operation of an adaptation is its *function*." They further proposed that "characters, evolved for other usages (or for no function at all), and later 'coopted' for their current role, be called *exaptations*" and that "the operation of a useful character not built by selection for its current role [is] an *effect*." Exaptation has been a very successful meme, but what if most features of organisms are modified versions of earlier features that once performed other roles? This would mean that there are almost no real adaptations and very few functions but mostly exaptations and their effects. Not only would adaptation fail to signify many interesting features, but this terminology would fail to distinguish between those *effects* of an exaptation that were beneficial and those that were detrimental or neutral.

The search for the "original" function of a conserved feature resembles the search for the "original" meaning of a folktale that has been retold many times and has meant many things to different listeners and different storytellers. There may indeed have been an original telling that was not a retelling of yet older tales—and recovery of this original sense would undoubtedly be of antiquarian interest—but the existence of an original sense should not belittle the subsequent senses in which the tale has been told and interpreted. Another example will drive home this point. A word may have an "original" meaning that is listed as *obsolete* in the *Oxford English Dictionary* and derived meanings that are currently in use. A pedant might insist that only the earliest definition is the "true" meaning and all subsequent usages are simply exaptations of this original sense; but the reason *why* the word is currently used is its role in the language as understood by contemporary speakers. Meaning is synchronic use.

A still-current sense of *dower* is "the portion of a deceased husband's estate which the law allows to his widow for her life," and this was also the "original" (now obsolete) sense of *dowry*. *Dower* and *dowry* appear to have been minor variants of the "same" word derived from Old French *douaire*, but a dowry is now "the money or property the wife brings her husband." *Dower* and *dowry* are currently understood to be "different" words. How did an obligation of the husband to the wife become a payment from the bride to the bridegroom? I do not know the answer to this question but can offer a just-so story to be tested against the historical evidence. In far-off days, O Best Beloved, dower was promised by the husband to provide for the wife in the event of his death. Highly desirable husbands then asked in their marriage settlements that the bride's family endow the property that would constitute the dower if the husband predeceased the wife.

Henceforward, what had been an obligation of the husband for support of the wife became a payment to the husband for taking the wife.

The essential oils of *Ocimum basilicum* (basil) are insecticidal compounds that evolved as a defense against insects. Human foragers would have moved through the woods nibbling the tip of a leaf here, and another tip there, and harvesting those plants that appealed to their palate. Palatable plants were therefore less likely to set seed. Appeal to the human palate was a maladaptation, not a purpose of wild basil, but the situation reversed immediately when foragers started collecting seeds from flavorful and fragrant plants and cultivating them. Now flavor was the reason for being planted. Seeds from less palatable plants were less likely to be planted and less likely to be protected from insects by gardeners. Seeds from plants that tasted good were more likely to be preserved. Appeal to the human palate had become a *reason for the existence* of basil plants.

What is the function of the essential oils of domesticated basil? One possible answer would be that their function is protection against insects and that appeal to the human palate is not a function but a serendipitous side effect. Another possible answer is that their function *was* protection against insects but now *is* appeal to the human palate. A choice between these answers is not a matter of identifying the *true* function; rather the choice indicates what the answerer *means* by "function." I prefer the second answer, but you are not wrong to choose the first; we simply mean different things by "function." A third answer could be that the function of the essential oils is both protection and flavor because the insecticidal function did not disappear with cultivation: cooks prefer to harvest leaves without chunks bitten out by caterpillars.

My fable of wild and domesticated basil is a retelling of Daniel
Dennett's (1987, 290) allegory of the wandering two-bitser. This
was a soft-drink vending machine that had been designed to
accept US quarter-dollars and dispense bottles in response. The
two-bitser was then transported to Panama where it was used
to dispense bottles when customers inserted quarter-balboas.
The two-bitser functioned well in this new task because quarter-
balboas were struck by the US mint from the same blank as US
quarter-dollars, and the machine could not distinguish between
the two currencies. Dennett considered the two-bitser's function was now to accept quarter-balboas even though it had been
designed to accept quarter-dollars, just as the function of the
essential oils of domestic basil is to appeal to the human palate
even though the oils had evolved to deter insects. With respect
to its current functions, the two-bitser would "malfunction" if it
dispensed a bottle in response to a US quarter in Panama or to a
lead slug that had been shaped to mimic a quarter-balboa (or any
other counterfeit coin). Similarly, the "designs" of natural selection can malfunction because the environment has changed or
because the adaptation enhances fitness in some but not all of
its current environments.

The word "function" is used in different senses in different stories. During the twentieth century, "functional morphologists"
developed a concept of function that was intended to exclude
considerations of end-directedness. Their research program was
strongly influenced by engineering with an emphasis on the
mechanical properties of Darwin's springs, wheels, and pulleys
but only a peripheral interest in the specific uses for which the
mechanical parts were employed. Theirs was a Baconian function
shorn of Darwinian purpose. Walter Bock and Gerd von Wahlert

(1965) exemplify this tradition. They championed a definition of function that did not invoke "any aspect of purpose, design, or end-directedness" (274). The function of a feature was "its action or how it works" and included "all physical and chemical properties arising from [the feature's] form" (273). The heart's functions included thumping sounds as well as the propulsion of blood. Form and function were seen as complementary aspects of physical features. In this formalism, the functions of a feature were simply all of its effects considered without regard for the biological role that the feature might serve. Thus, "the legs of a rabbit have the function of locomotion—either walking, hopping, or running—but the biological roles of the faculty may be to escape from a predator, to move toward a source of food, to move to a favorable habitat, to move about in search of a mate, and so forth" (279).

Ron Amundson and George Lauder (1994) defended a "causal role" concept of function that was "both non-historical and non-purposive in its applications" (466). They saw the goal of functional analysis as explaining "a *capacity* of a system by appealing to the capacities of the system's component parts" (447). Functions were capacities of particular interest (not simply all physical and chemical properties of a feature). The thumping of the heart was a simple and uninteresting property. "Scientists choose capacities which they feel are worthy of functional analysis, and then try to devise accounts of how those capacities arise from interactions among (capacities) of component parts" (447). By such criteria of functional *worth*, "Functional anatomists typically choose to analyze integrated character complexes which have significant biological roles" (450). For example, an anatomist might analyze the crushing capacity of a piscine jaw, but "While the decision to analyze the jaw may have been motivated by a knowledge of its biological role (the fish eats snails), that knowledge plays no part in the analysis itself" (451). Considerations of biological purpose could be used to choose which capacities were worthy of analysis, but these considerations were excluded from the analysis itself. Amundson and Lauder were thus able to choose objects of study by intuitive criteria of value but avow that their causal-role functions were uncontaminated by teleology.

Robert Cummins similarly rejected the grounding of function by natural selection and recent utility:

While it is plausible to suppose that there was a first flight-enabling wing somewhere among the ancestors of today's sparrows, those ancestors were not sparrows. . . . Similarly, somewhere in our ancestral lineage is to be found the first appearance of centralized blood circulation. But those ancestors were not even vertebrates. . . . It follows from these considerations that sparrow wings and human hearts were not selected because of their functions. Selection requires variation, and there was no variation in function in the structures in question, only variation in how well their functions were performed. (2002, 164–165)

But variation of function was not absent. Cummins ignores the ever-present background of mutational variation and negative selection. Mutation is an expression of the universal tendency to disorder. Negative selection maintains useful order. Some sparrow chicks hatch without functional wings. Some human embryos are conceived without hearts. Neither wingless chicks nor heartless embryos leave descendants because sparrow wings are needed for flight and human hearts are needed for the circulation of blood.

A complete history of everything that happened (the Laplacean *how come?*) is unattainable and unusable. A useful history needs to identify patterns of events of particular significance (the *reasons why* things are as they are). The evolving sequences of genes preserve a textual record of past differences that have made a difference. The previous chapter analogized this textual record with the Aristotelian notion of formal cause. The initial dissemination of fortuitous slips of the pen is the role of positive selection, but any repeatedly recopied text will accumulate errors if not corrected for sense (this is the role of negative selection). The chapter also provided a justification for ascribing purposes to the products of natural selection. Final causes (functions) were identified with the reasons why some textual variants have been favored over others. These reasons may change with time. One might give a different answer to why this variant has been favored over the last hundred generations or the last thousand generations, and a different answer for why it was favored a thousand years ago and a million years ago. That is the nature of historical explanation. The restriction of the laurels of function to the original reasons for positive selection ignores the importance of negative selection. A long history of negative selection is essential if any record is to survive of ancient positive selection.

The next chapter addresses the intentionality of designed or evolved automata that interact meaningfully with their world. It will propose that the meaning of information for an automaton be considered the automaton's output in response to the informative input. Some automata are simple: the meaning *for the two-bitser* of the "coin" that is inserted into its inner mechanism is whether or not a bottle is dispensed. Other automata are highly sophisticated, even capable of arguing about the meaning of "meaning." Dennett (1987, 387) compared organisms to highly sophisticated automata that derived their original

intentionality from natural selection but acted autonomously
in the world: "We, the reason-representers, the self-representers,
are a late and specialized product. What this representation of
our reasons gives us is foresight: the real-time anticipatory power
that Mother Nature wholly lacks. . . . We may call our own intentionality real, but we must recognize that it is derived from the
intentionality of natural selection, which is just as real—but just
less easily discerned because of the vast difference in time scale
and size."

In principio erat finis.

A mechanical device strikes a match and a candle is lit or there is an explosion. The striking of the match (+M) and the presence of oxygen (+O) are the same in both scenarios. The difference that makes the difference as to whether or not there is an explosion (±E) is the presence or absence of hydrogen (±H). A more sophisticated device strikes a match contingent on input from a hydrogen sensor. If the sensor fails to detect hydrogen (–H), the match is struck (+M). If the sensor detects hydrogen (+H), the match is not struck (–M). The first device is an effector of an explosion in the presence of hydrogen but does not "choose" the explosion because it does not "use" information. It couples a state of the world (±H) to an outcome (±E). The second device "prefers" darkness to an explosion in the presence of hydrogen. It couples one bit of information about the world (±H) to one degree of freedom in action (±M). It "responds" to +H with –M and to –H with +M. It is "undecided" until observation of what was uncertain (information) is interpreted in definite action (meaning).

The first difference in the phrase "a difference that makes a difference" is cause, the difference-maker or independent variable, and the second is effect, the difference-made or dependent variable. But whether the second difference is an interpretation of the first depends on the evolved or designed function of an interpreter. The first device does not interpret. Things just happen (±H, ±E). For the second device, the first difference is information (±H) and the second is meaning (±M). For an outside interpreter of the second device, ±H and ±M contain mutual (or redundant) information: either can be inferred to "mean the other." The observer can predict whether or not the match will be struck (meaning) by observing whether or not hydrogen is present (information) or could infer whether or not hydrogen was present (meaning) by observing whether or not the match was struck (information).

On what reasoned grounds can I claim that the second device interprets the presence of hydrogen as a reason not to light a candle but reject the seemingly parallel claim that the first device interprets the presence of hydrogen as a reason to cause an explosion? My argument makes an implicit appeal to what Ruth Millikan (1989) has called *proper functions*. The proper function of the first device is lighting candles. Explosions are unintended consequences. The proper function of the second device is the use of information to decide whether or not to strike a match.

The responses of an interpreter with one degree of freedom of interpretation may seem mechanical and uncomprehending, hardly deserving the "meaningful" label, but any truly sophisticated interpreter will have many degrees of freedom with multiple levels of internal interpretation in which the interpretation of one part of the system is news (and hence information) to other parts of the system. Consider multiple rewirings of a complex device such that each new device responds to the same inputs with different outputs. There is only a shallow sense in which the inputs of these devices are *causes* of the devices' outputs. An understanding of *how* inputs are interpreted as outputs requires an understanding of a device's inner workings. An understanding of *why* a device interprets particular inputs as particular outputs requires an understanding of the device's function and history. We cannot invoke an omniscient homunculus within the system that has an overview of what the system "knows" (Dennett 1991). An interpreter cannot "know" what it will choose until it chooses—if it "knows" it has chosen—but an observer can often predict a consistent interpreter's choices with confidence.

John Dewey recognized that "stimulus and response are not distinctions of existence, but teleological distinctions, that is, distinctions of function, or part played, with reference to reaching or maintaining an end" (1896, 365). A response implies a purpose. One cannot simply draw an arbitrary boundary around part of a complex web of processes, and then describe all causes crossing the boundary from outside to inside as stimuli, all causes crossing from inside to outside as responses, and all processes within the boundary as interpretation. Interpreters are intentional mechanisms that have evolved or been designed to use information in choice. Under this version of behaviorism it is not the input that determines the output but the relation between input and output that determines the mechanism.

Teleology of Interpretation

Meaning and *function* are intentional terms. The previous chapter grounded talk of biological function in the unintended teleology of adaptation by natural selection (following Dennett

1987, 1995; Millikan 1989; Neander 1991; Papineau 1984). Final causes were presented as efficient, even indispensable summaries of complex concatenations of efficient causes. A token effect cannot precede its token cause. But when one generalizes from causes of tokens to causes of kinds, effect-tokens both precede and succeed cause-tokens in recursive processes. A full causal account of an egg or a chicken contains long series of past chickens and past eggs. An egg is both an effect of a chicken-that-was and a cause of a chicken-to-be.

Natural selection subsumes all processes by which the environment *selects* a subset *from a set of actual things*. Reproduction replenishes numbers of the diminished subset before the next round of selection. Although nature's "choices" are unintended, some of her "choices" leave genetic records that allow repetition of that which was "chosen" as intended choices of living beings. Reproductive recursion is rescued from eternal recurrence of the same by the input of new variation via mutation and the shuffling of genetic texts in sexual reproduction. By these processes, recursively selected subsets accumulate information about what worked in the past. And what worked was the interpretation of information from the environment in "real time" rather than evolutionary time. As a consequence, the world now abounds with biological interpreters that *select* which differences will make a difference from the myriad potential causes in their environment and *choose* actions *from sets of possible actions* on the basis of observations that could have been different. The mapping of possible inputs to outputs is embodied in the interpreter's fine structure, with the fit between information and meaning—the efficaciousness of interpretation—derived from past natural selection refined by developmental processes during the interpreter's life.

To "intend" is to choose for anticipated effects. We can distinguish two kinds of intentionality. *Primary intentionality* is the repetition of causes that worked in the past. This is the intentionality of adaptation by natural selection and of conditioned reflexes. Past effects are anticipated to occur again. *Secondary intentionality* is choice of action after simulation of possible choices and their effects. Simulated effects are anticipated to occur when the action is performed. Secondary intentionality requires imagination, an ability to "hold in mind" and evaluate virtual outcomes. Primary intentionality is "primary" in the sense that anticipation evolved before imagination.

Before the Interlude, I wrote that information has meaning *for an interpreter* when it is used *to achieve an end*. The present chapter simplifies and clarifies by explicitly equating meaning and interpretation. The action or thing chosen is the meaning *of* the observation *for* the interpreter. Information resides in differences among things that remain "possible" until observation of an actual thing. Meaning is the response of the interpreter to the observation and is itself an actual thing that can be observed and used as information by another interpreter. By these definitions, "semantic information" is a contradiction in terms.

Information and Meaning

An interpreter can be viewed as an input–output device that uses observations to choose actions (figure 12.1). *Interpretation* subsumes all internal processes that couple observations (information) to actions (meaning).[1] The number of independent

1. Pearl's (2000) "causal models" can be considered a class of interpreters in which the *do*-operator performs the role of an observation by fixing

Figure 12.1
An interpreter is a computational mechanism for which information is
input and meaning is output.

things a device could observe can be considered a measure of its
uncertainty (entropy of observation). The number of independent
actions in its repertoire of response can be considered a measure of its *indecision* (entropy of action). Uncertainty is resolved
by observation, indecision by choice. Uncertainty and indecision are measures of potential things. Observations and meanings are actual things. The same observation can mean different
things to different interpreters, and different observations can
mean the same thing (i.e., be interpreted as the same choice of
action).

An interpreter's possible inputs are *the things to which it could
respond*. Its possible outputs are *the ways in which it could respond*.
These capabilities are competences of the interpreter, not properties of its world. They are subjective rather than objective.
Observations inform, whether the thing observed is ontologi-
cally uncertain (undetermined until observation) or epistemically uncertain (determined but "unknown" until observation).
Observation of what is epistemically uncertain provides information about prior events. Meanings may be "mistaken"
because of malfunction, unanticipated inputs, or because what
was once adaptive is now maladaptive. Unintended meanings

the value of an input variable. Tononi's (2004) "integration of information" refers to internal causal processes of interpreters.

may be used as information by other interpreters or by the same interpreter in self-reflection.

My purpose in this "behaviorist" account is not to belittle the complexities of interpretation but to argue that there is no ghost in the semantics. Information resides in distinguishable things in the interpreter's world, and the meaning of a particular input for the interpreter is simply whatever physical thing (print on paper, sound vibrations, neural states, etc.) is the output of information processing by the interpreter. The complexities reside in how the inner workings of the interpreter map observations onto actions. There is no nonmaterial domain in which meanings reside outside of physical interpretation. If you protest that this paragraph means more than ink marks on paper or pixels on a screen, then those ink marks or pixels have been input to a very sophisticated interpreter—your good self, and I thank you for reading.

Consider a paradigmatic small dark something moving across a frog's visual field that causes the frog to stick out its tongue to intercept the thing. If we treat the frog as a black box, photons falling on retinas are information (input); sticking out the tongue is meaning (output). If we were to peek inside the frog's brain, we would find multiple interpretations of interpretations between sensory excitation and motor action. My claim is that each physical state can be considered the meaning of prior information processing and that these states inform subsequent neural states that are themselves new meanings. The frog's visual system interprets incident photons as information about distance, direction of movement, and speed of the speck. These meanings inform subsequent interpretation as motor action. The frog minimizes immediate interpretation of a speck's nature so as not to give a fly time to interpret the frog's intentions (a small dark object in the mouth is worth ten flies that got away). Once the moving something has been intercepted, the frog has ample time to interpret whether the thing is food and what kind of food (using oral rather than visual sensors) and to modify its sensory criteria for future protrusions of the tongue.

Any spoken or written claim by a philosopher about what internal states mean to the frog—whether "fly," "food," or "small moving thing"—is an interpretation of the philosopher. It is thus the philosopher's meaning, not the frog's meaning. If we were to peek inside the black boxes of philosophers' minds there would undoubtedly be many interpretations of interpretations, meanings of meanings, before keys were struck on keyboards or words spoken or shouted. If a literate frog were to write a memoir of her experience, she might report that she saw the speck as a fly but was mistaken. Her interpretation would be of similar kind to the philosophers' interpretations. An interpreter, even an interpreter of itself, never has direct access to things in themselves but only to information about things.

The claim that meaning is whatever physical thing an interpreter interprets information to mean is a definition, not a claim that all interpretations are equally useful. Some interpretations are "better" than others because they inform more subsequent interpretations or enable meaningful interpretation of what was previously uninterpretable. Our perceptions have evolved to present useful information about the world to guide our actions, and our interpretations of words have evolved, from childhood, to make sense of what others are saying. Although information and meaning are defined relative to an interpreter *as subject*, interpreters may aspire or have evolved to interpret information *objectively* (Lindley 2000).

Interpretation of Interpretations

The nonliving world is a repository of unintended information useful for living interpreters. Unintended information is also present in interpretations of other interpreters. When an interpretation is reinterpreted, one must distinguish the intentions of the first interpreter (the producer) from those of the second interpreter (the consumer). The evasive movements of a gazelle pursued by a cheetah are intended to make the gazelle harder to catch. The cheetah observes and interprets these movements with the intention of catching the gazelle. These interpretations of the cheetah are unintended by the gazelle.

When a healthy gazelle sees a hunting dog rather than a cheetah, it interprets the situation as an occasion to stot (jump up and down). Hunting dogs preferentially chase gazelles that do not stot or stot more feebly. A vigorous gazelle and a hunting dog both benefit from the hunting dog chasing a feebler gazelle. The hunting dog's decision to chase a nonstotting gazelle is intended by the stotting gazelle. The evolutionary rationale of stotting is thought to be that stotting "signals" to the hunting dog that the gazelle has a good chance of outlasting the hunting dog in an extended pursuit and is therefore not worth chasing (FitzGibbon and Fanshawe 1988). But this is an interpretation of behavioral ecologists, not, as far as we know, of either gazelles or hunting dogs. Their interpretations are simply stotting and not chasing. (Gazelles do not stot to cheetahs because cheetahs lack endurance but are capable of short bursts of great speed. Gazelles interpret cheetahs as reasons to get far away as quickly and unpredictably as possible.)

Interpretations are actions chosen to achieve ends. Some interpretations are intended to be used as information by subsequent interpreters or by the interpreter itself at some later time. I will use *text* to refer to an interpretation intended to inform subsequent choice. A text is an output of an *author* (producer) intended to be input to a *reader* (consumer), but it is the reader who chooses how to interpret the text. A text anticipates interpretive competence of intended readers. It may be a static object or dynamic performance. By this expanded definition, written documents, works of art, DNA and mRNA, neural activity, and the tape of a Turing machine are all texts. My spoken words are an ephemeral text "written" in sound intended to be interpreted by listeners. A painting is a persistent text "written" in pigment intended to be interpreted by viewers. Parallel white lines crossing a road are texts intended to be interpreted by pedestrians as places to cross and by motorists as places where pedestrians cross. A peacock's tail is a text intended to arouse the admiration of peahens. Stotting is a text intended to discourage hunting dogs.

An author's intended interpretation of a text should be distinguished from the actual interpretation of the reader. A hunting dog who detects unintended evidence of weakness in a gazelle's performance may chase a stotting gazelle. The author's intentions are also distinct from how the author intends a text to be interpreted by readers. Some texts are deceptive. The folded wings of a camouflaged moth are intended to be interpreted as "not a moth" by moth-predators, but the flash of "eyespots" as the moth unfolds its wings for flight are intended to be interpreted, for a crucial split-second, as "eyes of a moth-predator predator." If these texts are interpreted by moth-predators as intended by the moth, then they have served their purpose and have been interpreted as intended by the author by being misinterpreted by the reader.

A completed nest contains clues about its construction. If birds model their own nest on the nest in which they hatched, then the parental nest informs the construction of the offspring nest. If parents constructed nests in ways that were easily interpreted by offspring and this method of construction was repeated because it enhanced the survival of the parents' grandoffspring in offsprings' nests, then the nest is a text of the parents with an intended interpretation by offspring. This example shows that a thing may function both as a tool (for holding eggs) and as a text (for instructing offspring). When the Mafia leave the body of an informer in a town square, the murder is both a direct means to an end (removal of an informer) and a text (a warning to potential informers).

The broken-wing display of a ground-nesting bird is a text intended to be interpreted as "here is easy prey" by a predator. The function of the bird's "helpless fluttering" is to lead the predator away from hidden eggs that are truly easy prey. If the predator recognizes the display as a text, concludes that "a nest is near," and searches for the nest, then the text has failed to be interpreted as intended but the predator has correctly interpreted the situation. If, on the other hand, the bird truly has a broken wing, then its seemingly helpless fluttering would indeed be helpless fluttering, and an attempt to escape, rather than a text with an intended interpretation. The predator would truly interpret the bird's actions if it concluded "here is easy prey." If the predator mistakenly interpreted helpless fluttering as "a broken-wing display" and started looking for a nest then the predator would have misinterpreted the situation.

One could imagine a bird with an actual broken wing "unconvincingly" attempting to escape with the intention that the predator interpret its actions as "this is a broken wing display;

therefore, the bird is not easy prey but a nest is near." If the predator started to search for a nest, then the predator would have recognized that the bird's actions were a text, would have interpreted the text as intended by the bird, but would have misunderstood the bird's intentions. The text would have achieved the bird's intention but foiled the intention of the predator who had been "deliberately" misled.

Ultraviolet photons damage unpigmented skin. Some skin responds to ultraviolet exposure by deposition of melanin. For skin that does not respond in this way, ultraviolet photons are not used as information and simply cause unintended damage. For skin that is able to darken, ultraviolet light is interpreted by the skin as melanin to prevent further damage. An observer of tanned skin can infer that the skin has been exposed to the sun and that, depending on context, the skin's owner works outdoors or has sufficient leisure to spend time at the beach. Because possession of leisure has social value, some people choose to expose their skin to solar lamps so as to be interpreted as possessors of leisure. In this case, tanned skin is a text intended to inform the interpretations of intended observers.

An interpretation (a meaning) can be considered to "represent" the interpreted information and to be an ascription of content to the information *by the interpreter*. (The interpretation stands as a metaphor of the observation.) Representation raises intriguing questions about how internal processes of an interpreter derive meaning from information and why the interpreter has evolved or been designed to interpret information in the way that it does. A text not only "represents" information used by its author in the text's composition but "presents" information to intended readers. Presentation raises additional questions of how authors anticipate readers' responses.

Texts act not directly but indirectly, through the interpretations of readers. Texts are not agents. They do nothing, in and of themselves; but one need only consider the Bhagavad Gita, the Declaration of Independence, or *The Protocols of the Elders of Zion* to appreciate the differences texts can make in the world.

Private and Public Texts

The inner workings of complex interpreters involve delegation of tasks among subinterpreters, each presenting texts for use by other subinterpreters. These private texts, intended for internal use, all have a physical form. Some are ephemeral. A percept is an interpretation of sensory input intended to inform subsequent interpretation. Others are persistent. A memory is a textual record to be consulted when needed. Consciousness is a private text, our mental desktop, a very short-term memory written in we know not what medium, consulted by other subinterpreters to know where to "look" for relevant information. The scene that is seen functions as a simplified and constantly updated interpretation of the world that is compared against incoming percepts for detection of relevant differences.

I view a landscape by Claude Monet in which artfully arranged splodges of color on canvas present a scene on the Seine with sailboats and a group of five ducks to one side. Not only do I see the play of light on water, I hear the clink of rigging as the boats rock at anchor. These meanings are my interpretations. My companion sees gulls rather than ducks. As I move in toward the canvas, the things that I saw dissolve into ill-formed blobs. The ducks are revealed as no more than five dabs of white paint. The scene that I saw was underdetermined by input from my eyes, but Monet's genius was to suggest, with minimal means, internal sources of information that filled in the picture. He created the illusion of detail that I expected to find when I looked closely. This paragraph does not contain all the information I want to inform your interpretation. By the artful placement of ink on paper I want to suggest paintings and scenes you have seen in order to evoke an "Aha! I see what he means." I particularly want you to see that an author always relies on rich sources and resources in the private texts of his readers for them to make sense of his public text.

My text is the product of multiple drafts of an evolving text. In the process of reading and rereading, writing and rewriting, I came to understand what I meant and mean. My meaning is the public text that you see, not some nebulous sense in my mind to which the text points. As my aging mind becomes less nimble, I rely more and more on public texts of previous selves as *aide-mémoire* of what I wish to mean. What persists in my brain are reworked memories of earlier drafts and regrets about what I wrote poorly. Once a text is published, its meanings for readers are untethered from its author's intentions. My meaning becomes your meaning as you read my text.

Meanings of Genes

What do genes mean? The short answer is whatever physical thing an interpreter interprets them to mean. Cells contain three very important interpreters that evolved long ago to interpret aperiodic polymers: *DNA polymerases* complement sense-strands of DNA with their antisense-strand; *RNA polymerases* transcribe sense-strands of DNA as RNA; ribosomes translate messenger RNAs (mRNAs) as proteins.

For a DNA polymerase, the meaning of a DNA strand is its complement. For an RNA polymerase, the meaning of a DNA strand is an mRNA. For a ribosome, the meaning of an mRNA is a protein. Thus, the same DNA sequence means different things to a DNA polymerase and an RNA polymerase. The information (input) is the same, but the meaning (output) is different. DNA and mRNA are texts intended to be interpreted. RNA polymerases also transcribe transfer RNAs (tRNAs) and ribosomal RNAs (rRNAs). These are tools to be used in translation of mRNAs, not texts to be further interpreted. Among the texts interpreted by RNA polymerases and ribosomes are instructions for the assembly of RNA polymerases and ribosomes. Interpreter know thyself.

These molecular machines are general-purpose, mindless interpreters of specialized texts. DNA polymerases, in particular, resemble monkish scribes in the scriptorium of the Total Library of all possible texts (Borges 2000), faithfully copying DNA sequences that are both unilluminated and unilluminating. One can posit a Library of Mendel that contains all possible DNA texts of some finite length (Dennett 1995). Only an infinitesimal subset of texts the length of the human genome can ever have existed. Natural selection has acted in this much more restricted, yet unimaginably vast Library of Darwin that contains all past and present DNA texts. In this library, differences between texts that are still read and texts that are no longer read convey information about what has and hasn't worked in the past.

A DNA strand is an interpretation of its antisense, via one round of DNA replication, and of its sense, via two rounds of DNA replication. But a gene can also be interpreted as an mRNA, via transcription. Thus, a gene means itself for a DNA polymerase but means an mRNA for an RNA polymerase. Complex interpreters can be built out of simpler interpreters. A DNA strand is interpreted as a protein by the combined interpretative system of an RNA polymerase and ribosome via two steps of interpretation, first transcribed as mRNA, then translated as protein. Thus, a gene means a protein for this compound interpreter.

A more contentious claim is that an organism's genes collectively mean the organism. Past organisms have been responsible for the replication and transmission of present genes, which are interpreted via complex processes of development to produce present organisms that are responsible for the replication and transmission of future genes. Organisms and their genomes are thus recursively related via primary intentionality. As such, organisms can be considered to interpret their genomes as themselves. This bald statement should not be overinterpreted. Every interpretation of a genomic text as an organism is unique because texts are always interpreted in the *context* of other sources of information. Organisms interpret their genomes in environmental context and not every detail is intended.

Minor genetic differences may have major organismal effects. The 282nd amino acid of human factor VIII proteins is arginine, specified by the codon CGC on the sense-strand of the *factor VIII* gene. The complementary codon on the antisense-strand GCG is the template for transcription of CGC in the mRNA which is translated as arginine by the ribosome. Cytosine (C) preceding guanine (G) can be chemically modified by attachment of a methyl group to produce 5-methylcytosine (5-meC). Spontaneous deamination of 5-meC converts 5-meC into thymine (T), creating a heritable mutation that changes antisense GCG to GTG, which is interpreted as sense CAC by DNA polymerase. RNA polymerase transcribes GTG (DNA) as CAC (mRNA), which is translated as histidine by the ribosome. Factor VIII protein, with histidine rather than arginine as the 282nd amino acid, fails to clot blood causing life-threatening hemophilia in males.

Mutations in DNA that change an amino acid in protein are described as *nonsynonymous*. DNA polymerases, RNA polymerases, and ribosomes faithfully interpret nonsynonymous mutations without regard for the functionality of the resulting protein. For a DNA polymerase, CAC means GTG (antisense) and CAC (sense); for an RNA polymerase, GTG means CAC; for a ribosome CAC means histidine. These are the intended meanings of interpreters that have evolved to represent whatever text they are presented. But, histidine and hemophilia are unintended from the perspective of the organism: none of the fetus's male ancestors possessed the mutant protein or suffered from hemophilia. A genetic counsellor reads a printout of the DNA sequence of the *factor VIII* gene present in amniotic fluid of a male fetus whose mother is a carrier of a mutant *factor VIII* gene on one of her X chromosomes. The difference that will make a difference is G versus A in the middle position of the 282nd codon of the *factor VIII* gene. Before viewing the printout, the counsellor is uncertain about what the data will show. If she reads G, then she can confidently inform the parents that the child will be unaffected; but if she reads A then she must inform them that the child will be affected.

Mutations in DNA that do not change the amino acid translated by ribosomes are described as *synonymous*. An evolutionary biologist might use the ratio of nonsynonymous to synonymous differences between two DNA sequences to infer whether the sequences have been subject to natural selection. In this case, meaningless differences for an intended interpreter (a ribosome) are meaningful differences for an unintended interpreter (an evolutionary biologist).

Telegraphing One's Intentions

> An engineering communication theory is just like a very proper and discreet girl at the telegraph office accepting your telegram. She pays no attention to the meaning, whether it be sad or joyous or embarrassing. But she must be prepared to deal intelligently with all messages that come to her desk.
>
> —Warren Weaver (1949)

The Zimmermann telegram of 1917 conveyed an offer of a military alliance between the German Empire and the Republic of Mexico if the United States of America entered the Great War. In the event of such an alliance, Germany promised to provide financial support to Mexico and recognize Mexican reconquest of Texas, New Mexico, and Arizona.

Great Britain cut all German telegraphic cables to the New World on August 5, 1914. Therefore, the German message to Mexico had to be conveyed by an indirect route. A plaintext of the message in German was composed in Berlin, under the direction of the secretary of state for foreign affairs (Arthur Zimmermann), and then encrypted using code 7500. The encoded message was passed to the American Embassy in Berlin for transmission by US diplomatic channels to the German Embassy in Washington. The message was sent by cable from Berlin to Copenhagen, from Copenhagen to London, and then by transatlantic cable from London to the State Department in Washington, DC. The encrypted message in code 7500 was passed to the German Embassy in Washington where it was decrypted into German and then re-encrypted in code 13040. Re-encryption was necessary because the German Embassy in Mexico lacked a code book for the more secure code 7500. The message in code

13040 was sent by telegram from the Western Union office in Washington to the Western Union office in Mexico City where it was printed as a typescript and delivered to the German Embassy in Mexico on January 1, 1917. The typescript was decrypted into German by the ambassador's secretary, before the offer was communicated by the ambassador to the Mexican President in Spanish.

Unbeknownst to the governments of the United States and Germany, the telegraphic signals that included the message in code 7500 had been intercepted by British intelligence in London and the message partially decrypted. The British then requested their agents in Mexico obtain a copy of the telegram in code 13040. The major reason for this step was to hide from the US government that British intelligence was reading their diplomatic cables. British intelligence decrypted the telegram in code 13040 (it contained little unexpected for them) and then the British government provided the telegram to the American government in its coded form and as versions decrypted into German and translated into English. The message was then released by the pro-war party in the United States to the American press. The popular outcry hastened the entry of the United States into the war on the side of Great Britain and its allies. The British government had decided that the benefit of revealing the telegram to the United States would outweigh the unavoidable information provided to Germany that Britain could read code 13040.

The Zimmermann telegram passed through many texts in many media, with interpretation occurring at each conversion from one medium to another. Sometimes "the message" was in German, sometimes in code 7500, sometimes in code 13040, sometimes in Morse, sometimes in Spanish, and sometimes in English. For the various clerks in telegraph offices in Berlin, Copenhagen, London, Washington, and Mexico City, the meaning of the text was simply the "mechanical" transcription of a stream of dots and dashes into an otherwise uninterpretable string of numbers and spaces or the equally mechanical conversion of a string of numbers and spaces into a sequence of dots and dashes. These clerks were employed to interpret all texts mechanically and lacked the context to interpret encrypted texts in any other way. The receipt of an encrypted text by a clerk conveyed a very large amount of information, because the clerk had few expectations about what the text might be, but the encrypted text had no intelligible meaning for the clerk who lacked the key to decrypt the message.

The staff of the German embassy in Washington obtained less information from receipt of a typescript in code 7500 than did the clerks through whose hands the message had passed, because the staff had much stronger expectations about the message. For example, the embassy staff expected the message to consist of numbers and spaces only, whereas the absence of letters was not a prior expectation of the telegraph clerks. But the message had much richer meaning for the embassy staff because they possessed code book 7500 and were therefore able to interpret the message as words in German to be re-encrypted in code 13040 using a different key. Because the staff were fluent in German, they undoubtedly formed a mental interpretation (a memory) of the telegram's content that could be called upon in the subsequent investigation by German intelligence of how the message had been intercepted.

Now consider the communication of the message to Mexican president Venustiano Carranza. The German ambassador to Mexico, Heinrich von Eckardt, possessed a typescript (perhaps a manuscript) in German from which he made a verbal offer to Carranza in Spanish. Eckardt's interpretation of ink marks on paper as words in German involved complex interpretation into a neural text that underwent an equally complex process of translation into Spanish in the speech-production centers of his brain. The neural text from the speech production centers was then conveyed to Eckardt's vocal apparatus, where it was composed as a phonic text (sound vibrations) that was received and interpreted by Carranza's ears and reinterpreted at various levels in Carranza's brain. At every step in the process, the constantly changing meaning was nothing more than the physical text produced by an interpreter, whether this consisted of ink marks on paper, neural representations in brains (whatever thoughts may be), or vibrations in the air.

Information resides not in things but in the reduction of uncertainty (entropy) of an interpreter who observes a thing. That which is known does not inform. If we suppose that American intelligence kept copies of German diplomatic telegrams sent from the Western Union office in Washington, then the American government learned nothing new from the Mexican telegram provided to them by the British government beyond that the British had a copy. What provided new information was the decryption of the telegram by British intelligence into written texts in German and English. American uncertainty about whether these texts contained British disinformation was dispelled when Zimmermann conceded that he had sent the telegram and that the decryption was accurate.

The coded texts of the Zimmermann telegram were intended to be uninterpretable (meaningless) for unintended readers. This was the only reason Germany would have sent such a message via the intermediary of the US State Department. The various code books were the private keys intended to be used by intended readers to interpret the strings of numbers and spaces of a typescript as an intended text in German. British intelligence sought to reconstruct this intended interpretation without possession of the private key. Unintended information in the arrangement of code groups ("similar" words had "similar" codes) provided clues that allowed British intelligence to construct their own key and interpret the message. The task of British intelligence was made easier by an unintended difference: Germany used numeric codes, the United States used alphabetic codes. German interpolations in the American cyphertext stood out like a sore thumb.

My account of the convoluted transmission of the Zimmermann telegram is reconstructed from Friedman and Mendelsohn (1994), von zur Gathen (2006), Freeman (2006), and Boghardt (2012). My interpretation of what happened is made difficult by intentional disinformation and unintentional misinformation in the historical record. And I may have misinterpreted the texts I have read and so may have unintentionally misinformed you about the details. My words will nevertheless have served their purpose if your interpretation of what I have written approximates how I intend you to interpret my text.

Mutual Information and Meaning

The concept of "semantic information" views meaning as present in information prior to its interpretation. On this view, interpreters repackage preexisting meaning in new media. I propose instead that meaning be considered the output of the interpretive process of which information is the input. On this view, answers to questions of meaning should be sought in the study of mechanisms of interpretation and the origins of interpretive competence. Once these difficult questions are answered, there will be no remaining ghost in the semantics. A preference between "semantic information" and "meaning as interpretation" is a choice of how "meaning" is defined, not a judgment of fact. "Meaning as interpretation" sits comfortably with the belief that information is not an objective property of things in the world but represents the epistemic uncertainty of an observer.

Consider two sequential transformations of Arthur Zimmermann's message to Heinrich von Eckardt (figure 12.2). In the first, Zimmermann's plaintext was translated into cyphertext by one of his secretaries using the textual prosthesis of codebook 7500. In the second, the cyphertext was re-encoded as an electrical signal by a telegraph clerk using the mechanical prosthesis of a transmitter. If these transformations were performed as intended, then the cyphertext could be reconstructed from the signal by a telegraph receiver and the plaintext could be reconstructed from the cyphertext by a reader in possession of codebook 7500.

Faithful transmission of the "Zimmermann telegram" depended on *mutual information* (statistical dependencies) among multiple coding systems that allowed messages to be encoded and decoded by senders and receivers with the appropriate codebooks. Proponents of "semantic information" often equate meaning with mutual information or consider meaning to be derived

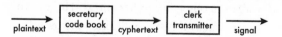

Figure 12.2
The interpretation of the plaintext of the Zimmermann telegram as a telegraphic signal.

from mutual information. In their view, the various texts of the "Zimmermann telegram" are vehicles of a common meaning conserved through the many transformations of transmission. From the perspective of "meaning as interpretation," mutual information allows an observation of one thing to be *usefully* interpreted as *about* something else. Meaning resides not in the immediate observation but in the interpretive synthesis of observation with background information (context), in order to act effectively in the world.

The world contains many unrecognized associations between values of one thing and values of another. These associations are "recognized" once an interpreter embodies a mechanism that couples observation of one to useful action on the other. Such mechanisms can be conceptualized as a codebook (text) that enables meaning to be "read" from observation. Codebooks codify mutual information in useable form. The intended extension of "codebook" in this paragraph includes all forms of embodied knowledge revealed in effective action. (Consider the physical implementation of the "genetic code," which is represented nowhere in tabular form except in recent texts of biologists.) Under this general definition, "codebook" and "interpreter" become synonyms. There are two broad ways of obtaining a codebook. The simplest is to be given a copy (or to steal one). By this means, copies of codebook 7500 facilitated secret communication between the Foreign Office in Berlin and the German Embassy in Washington. The second is to construct a codebook by statistical *inference* from observations in context. This was the more difficult task achieved by British intelligence and by a child as she learns a language.

A definition of meaning as interpretation simplifies many semantic problems. These problems include polysemy, how the same observation can have different meanings for different interpreters; synthesis, how information from multiple sources can be combined to generate new meanings; and the status of perverse but sincere interpretations. Consider divination by tea leaves. A fortune-teller observes the configuration of leaves in the bottom of a teacup and then uses the observed pattern to answer a question posed by a client. The pattern informs the fortune-teller's advice if a different pattern would have resulted in different advice. Do all possible interpretations exist as "semantic information" in the configuration of the tea leaves, or does the fortune-teller mistakenly believe she has identified "semantic information" that is absent? The redefinition of meaning as output avoids the horns of this dilemma and the need to distinguish "true" from "false" meanings. If different fortune-tellers (or the same fortune-teller at different times) interpret similar patterns as similar fortunes, then mutual information exists between tea leaves and fortunes; but these are meanings of the fortune-tellers, not of the tea leaves.

Information Theory and Meaning

> The fundamental problem of communication is that of reproducing at one point either exactly or approximately a message selected at another point. Frequently the messages have meaning; that is they refer to or are correlated according to some system with certain physical or conceptual entities. These semantic aspects of communication are irrelevant to the engineering problem.
> —Claude Shannon (1948)

Claude Shannon (1948) and Warren Weaver (1949) developed information theory as part of a broader theory of communication. Weaver conceptualized communication in "a very broad sense to include all of the procedures by which one mind can affect another" (1949, 11). He recognized technical, semantic, and influential problems of communication: technical problems were "concerned with the accuracy of transference of information from sender to receiver"; semantic problems were "concerned with the interpretation of meaning by the receiver, as compared with the intended meaning of the sender"; influential problems were "concerned with the success with which the meaning conveyed to the receiver leads to the desired conduct on his part." Thus, Weaver used "information" in the context of technical problems but "meaning" in the context of semantic and influential problems. Shannon's "fundamental problem of communication" was the technical problem of reproducing a message accurately at another place. He adopted the design stance to address this technical problem, and his mathematical theory ignored problems of intentionality. Weaver, by contrast, adopted the intentional stance in talking about the semantic and influential problems ("intended meaning," "desired conduct"), but these problems eluded mathematical treatment (see Dennett 1987 for discussion of the design and intentional stances).

In Shannon's theory "the actual message [was] one *selected from a set* of possible messages" (1948, 31). He used logarithmic entropy—the number of independent ways in which the message could have been different—as a measure of information content. "If the number of messages in the set is finite then this number or any monotonic function of this number can be regarded as a measure of the information produced when one message is chosen from the set, all choices being equally likely" (379). For Weaver, this meant that "the word information relates not so much to what you *do* say, as to what you *could* say. That is, information is a measure of your freedom of choice when you select a message" (1949, 12). Information measured in this way had "nothing to do with meaning." For Shannon, a message possessed *meaning* if it referred to, or was correlated with, physical or conceptual entities "according to some system" (379). This conceptualization of meaning had two components: the first was the correlation between messages and things in the world (mutual information), and the second was the "system" that embodied the correlation.

Figure 12.3 is a variant of a famous figure from Shannon (1948) in which a message is selected by an information source and passed to a transmitter that translates the message into a signal that is transmitted to a receiver that back-translates the received signal as a message that is passed on to the destination. The signal sent and signal received differ because of inputs of unintended information (noise). The central focus of the figure is the channel between transmitter and receiver. The technical challenge was to ensure that the message passed to the destination by the receiver was as close as possible to the message passed to the transmitter by the information source. In Weaver's example, "When I talk to you, my brain is the information source, yours the destination; my vocal system is the transmitter, and your ear with the eighth nerve is the receiver" (1949,

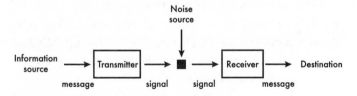

Figure 12.3
Shannon's (1948) diagram in which a message selected by an information source is transmitted to a destination.

12). Weaver and Shannon had relatively little to say about how information sources selected a particular message or about how destinations interpreted the message after its receipt (nor did they have much to say about the internal workings of transmitters and receivers). In terms of my interpretation of meaning, information source, transmitter, receiver, and destination are all interpreters.

Weaver and Shannon recognized the *interpretation of information* to be an important question, but their focus was on the *transfer of information* between interpreters. This chapter addresses how information is used rather than how it is transmitted and is concerned with the general problem of interpretation (use of information) of which communication—the production and interpretation of texts (things intended to be interpreted)—is a special case. The domain of interpretation includes the use of unintended information from the environment in addition to the interpretation of texts.

Difference Making and Mechanisms

Preference is an attribute of relations (differences and samenesses) rather than things (objects and events). To be told that someone chose x tells us nothing about his preferences unless we are also told what he rejected. When we choose one thing rather than another, we do not unconditionally endorse the thing chosen but express a preference for that thing relative to the other. We might be "making the best of a bad bargain" or "choosing the lesser of two evils."

A dichotomous choice between x_1 and x_2 can be usefully expressed as $\bar{x} \pm \Delta$ where $\bar{x} = (x_1 + x_2)/2$ is a sameness and $\Delta = (x_1 - x_2)/2$ is a difference. Being told the relations of things, $\bar{x} \pm \Delta$

conveys the same information as being told their identities, x_1 and x_2, but the relational form clarifies what is at stake. We choose from what is different ($\pm\Delta$) *in the context* of what is the same ($\bar{x}$). At the moment of choice $\bar{x}$ is the point to which we have come and $\pm\Delta$ a divergence of possible futures: a future to live and a future to die. Once a choice has been made, the future-chosen is past, the latest turn in the history of paths taken, and the future-rejected is what could have been. *How come?*—the path taken—preserves a trace of *Why not?*—the reasons for the rejection of paths not taken (if reasons there were).

Ponder again the flame that causes or does not cause an explosion. On one hand, the striking of the match and the presence of oxygen do not make a difference. It is the presence versus absence of hydrogen that makes the difference. On the other hand, the striking of the match and the presence of oxygen are essential parts of the mechanism that causes the explosion. When a scientist performs a controlled experiment, with and without hydrogen, she converts a possible difference into two actual series of events. (If she varied the presence of oxygen or striking of the match, then these become experimental variables and potential difference makers.) Observations are of actual things, not differences among possible things. Actual things, not differences, participate in mechanisms. But we study mechanisms to make a difference.

Causation as difference making is what could have been different. It is a history of paths not taken. Causation as mechanism is what could not have been otherwise. It is one damn thing after another, a single path. The relation between these two concepts of cause is much debated (Hall 2004; Strevens 2013; Waters 2007). Different actions could have different outcomes, but particular actions cause particular outcomes. We readily interpret the world as identities or relations and slip facilely between these perspectives.

An interpreter is an evolved or designed mechanism that couples *possible* inputs (entropy of observation) to *possible* outputs (entropy of action). These degrees of freedom are capabilities of the mechanism (what it *could* observe and what it *could* do). The interpreter is uncertain and undecided until an *actual* input is interpreted as *actual* action. Information is what could have been otherwise before observation. Meaning is what would have been otherwise had the observation been different. Interpretation couples information (difference maker) to meaning (difference made). For an interpreter that has evolved to intervene in its own fate, the only useful information is about differences that could make a difference.

Causation as difference making projects our epistemic uncertainty onto our mechanisms of choice. In a singular universe in which token events happen once, why should "could have been otherwise" be less respectable than "could not have been otherwise"? What difference does it make? How can one decide? Fisher (1934) flipped a bit in my mind, and I now choose "could have been otherwise," but I understand neither the difference nor what is at stake. The bit *could* flip back. From the perspective of "could have been otherwise," our choices *change* the world. We have evolved to choose because past choices have made a difference and future choices *will* make a difference.

The Parable of the Bathtub

> If you make yourself really small, you can externalize virtually everything.
> —Daniel Dennett (1984)

The study of mechanisms is frequently connected with reductionism, the ideology that larger things are properly explained by the properties and interactions of their smaller parts. It is incontrovertible that events at a small scale can have large effects. A mutation in a *factor VIII* gene of a single cell resulted in at least ten of Queen Victoria's male descendants suffering from hemophilia, including heirs to the thrones of Russia and Spain (Ingram 1976). All genetic differences that distinguish armadillos from aardvarks and zebras from zebus originated, in this way, as unintended mutations in single cells.

It should also be incontrovertible that large things affect small things. My paternal grandfather was gassed in the Great War. As he sat upright for inspection in his hospital bed, he probably raised his right hand to his head. (He performed this action when dying sixty years later in a different hospital bed.) The origin of the hand salute goes back centuries, perhaps millennia, lost in the fog of time. Among many theories, the salute is the ritualized removal of a hat in the presence of a superior, the raising of the visor by a knight in armor, or a demonstration that the weapon hand is empty, but this is conjecture. John Stewart Haig's salute would have been an almost automatic gesture in the presence of an Australian or British officer but would not have been elicited by other outwardly similar persons; definitely not by a German officer. From a mechanistic perspective, his salute was caused by release of acetylcholine at neuromuscular junctions triggering actin filaments to slide past myosin filaments in muscle fibrils of his arm. But what molecular mechanism caused his hand to be raised in the presence of a person of an abstract kind? How does a military tradition move an ion across a membrane?

The detritus from collisions between protons moving in opposite directions at $0.99999999c$ with a combined energy of

13 TeV has been interpreted as evidence for the existence of the Higgs boson. Such events are created and detected at the Large Hadron Collider, housed in a circular tunnel 27 kilometers in circumference, spanning the Swiss–French border. The even more powerful Superconducting Supercollider was canceled by the United States Congress in 1983 because of its immense cost. The differences that made the difference between the Higgs boson being detected in Europe rather than North America are to be sought in alternative political arrangements. No interpretable subatomic mechanism explains why two protons collide at these enormous energies in France but not in Texas. Events at the level of interpersonal relations and trans-Atlantic rivalries impinge upon, and predict, the movements of fundamental particles. Choices are the means by which big complex things control small simple things to make big differences.

An event's causal influence can wax or wane. A "butterfly effect," whereby the flapping of wings in Marilia causes a downpour in Sydney, expresses the intuition that small differences can have large effects. The existence of dynamic attractors has the opposite implication. Attractors cancel differences. Consider a bathtub. Whether water enters the tub from a spilled cup, the shower head, the cold faucet, or the hot faucet, it goes down the drain. The tub is literally a basin of attraction. All the degrees of freedom of water molecules to move "as they will" come to naught because the predispositions of the tub are imposed upon them willy-nilly "against their will." Each water molecule has a unique narrative of how it came to the drain, but these disparate histories have no future consequences. These are differences that do not make a difference. It is the *form* of the tub that cancels past differences, not the *matter* from which it is made. The bathtub enforces its will.

The vortex that forms as water exits a tub is a recurrent attractor. The living world abounds in recurrent goal-directed attractors because "endless forms most beautiful and most wonderful have been, and are being, evolved" (Darwin 1859, 490). Convergent structures are attractors in adaptive space over evolutionary time. Adult forms are attractors in morphological space over developmental time. Innate and learned actions are attractors in performance space over behavioral time. Cultural conventions are attractors in social space and the means by which a group's members converge on common meanings. Minor variations of font or pronunciation make little difference in how words are understood. Word tokens are rocks of stability that facilitate freedom of expression. A salute is a strange attractor indeed.

Organisms embody elaborate hierarchies of homeostatic attractors that ensure basic functions are unperturbed by causal fluctuations at multiple spatial and temporal scales. Bodily attractors at all levels—from the molecular, to the cellular, to the individual—buffer organismal fates from the unforeseen. This cancelation of irrelevant differences allows "focus" on what is relevant. From the myriad potential causes in its world, an organism selects those causes that are applicable for its needs to intervene adaptively at decision points. The organism is an "unmoved mover" moved by self-selected information in pursuit of intended ends. It determines which differences will make a difference. The regress of responsibility stops here. Organisms pull their own strings.

The Meanings of Life

Life is eternal recurrence. That which hath been shall be and that which was done shall be done again. But when a sperm fuses with an egg, the nascent interpreter is something new under the sun. Our genes convey information from the deep past to be used with information from current events to inform our choices in a rapidly changing world. Why a zygote receives one set of genes rather than another can be considered chance, a concatenation of random events that picks one from the numberless possibilities that could have been sampled from its mother's and father's genomes. The cards are ancient, but each hand is new. The fall of the cards cannot be anticipated, but we attempt to do the best with the cards we are dealt. We play our hands and the cards are reshuffled.

The meaning of a life is the life that is lived. Your body is an interpretation of your life. Cartesian dualism divided the indivisible body into tools to be used once a choice had been made (*res extensa*) and parts that were texts used in choosing (*res cogitans*). When an organism acts, it is the organism's evolutionary and developmental history that determines to which information it responds and how it responds. The organism exerts its will as the lead actor in its own drama, as a canceler of irrelevant causes and of competing narratives. Yet, despite this autonomy of action, many actors succumb to the slings and arrows of outrageous fortune, to factors beyond their control, foremost among which are the roles played by other actors.

Self-reflective organisms (selves) respond to their world with internal changes that rewire connections between inputs and outputs to make more effective future choices, learn from experience which inputs to attend and which to ignore, perfect performance by practice with feedback from past action, and possess rich memories to inform future choice. Highly sophisticated selves augment their behaviors by observation of what works for other actors, learn from the instruction of parents and other tutors, and choose principles by which to live in pursuit of self-chosen goals. These internal changes comprise an embodied memory of the self's life experience (the meaning of its life). This intricate and intimate private text, an interpretation intended to be self-interpreted, seamlessly melds ancient wisdom of genetic and cultural texts with news from the senses. It is responsive and responsible. It is the material and mortal soul that dies with the body.

In the beginning was mechanism. Things just happened. The origin of meaning was the origin of intentional difference making. Choice became free as degrees of freedom of observation and of intended action broadened and deepened. To understand a free choice, one must understand an interpreter's soul.

In fine est principium.

X   *Vive la différance*

It is also in this sense that the contemporary biologist speaks of writing and pro-gram in relation to the most elementary processes of information within the living cell.

—Jacques Derrida (1976)

A gene is an evolving text. Let us represent it by X. Its most recent version, X_0 was a choice of nature in which $X_1 + \Delta_1$ was chosen and $X_1 - \Delta_1$ erased where X_1 was a prior sameness and $\pm\Delta_1$ a prior difference. (In mathematical parlance, Δ is the symbol of difference and $\pm$ the union of positive and negative in the interchange of sign.) The deconstruction of X is *iterative* because X_1 was itself a prior choice in which $X_2 + \Delta_2$ was chosen and $X_2 - \Delta_2$ erased. Thus, X_0 can be deconstructed as $(X_1 + \Delta_1)$, which can be deconstructed as $((X_2 + \Delta_2) + \Delta_1)$, which can be deconstructed as $((X_3 + \Delta_3) + \Delta_2) + \Delta_1)$ and so on to the nth degree of deconstruction:

$$((((((X_n + \Delta_n) + \Delta_{n-1}) + \Delta_{n-2}) + \Delta_{n-3} \ldots + \Delta_3) + \Delta_2) + \Delta_1).$$

At the nth-level of bracketing of the text, the originary text X_n is a minor term in the accumulation of differences and could itself be further deconstructed as a sum of earlier differences. The text is a blank, struck between hammer and anvil, that bears an impress of past choices ($\Sigma\Delta$) and erasures ($-\Sigma\Delta$) where Σ signifies summation. The trace of differance ($\pm\Sigma\Delta$) is the value of the coin. This ~~mathematical model~~ sentence is an undisciplined metaphor of meaning ~~of a gene~~, an evolving text that never comes to closure and is interpreted anew at each ~~writing~~ reading. ~~Choices and erasures are not values on a line~~ The tense of the text, the progressive present of *being* written, is poised between the past perfect of *has been* rewritten and future imperfect of *will be* rewritten. A sense of the text unfolds as *being* read.

Le champ de l'étant, avant d'être déterminé comme champ de présence, se structure selon les diverses possibilités—génétiques et structurales—de la trace. (Derrida 1967, 69)

The field of ~~the entity~~ being, before being determined as the field of presence, is structured according to the diverse possibilities—genetic and structural—of the trace. (Spivak translation of Derrida ~~1976~~ 2016, 51)

Deconstructing Derrida

In an early rewriting of the previous chapter, I adopted "text" to designate "an interpretation intended to be interpreted." I was vaguely aware that theories of the text were central to recent scholarship in the humanities, but this was a literature of which I was poorly informed. After my text stabilized, I borrowed a copy of *Of Grammatology* (Derrida 1976) to gain a sense of what its author had to say about texts. An early impression was that Derrida and I were writing about the *same* relations in *different* language. Were genes the originary arche-writing, the first durable institutions of signs? Could Derrida and Dawkins be on the same page with a shared concern for the centrality of inscription?

Derrida rejected the idea that we have immediate access to something exterior that is presented to consciousness as *being*.

He wrote of the writing and rewriting of texts. Daniel Dennett similarly rejected a Cartesian theater, the mythic place in the brain where experience is presented as consciousness. He wrote that there was no canonical text, just multiple drafts subject to endless revision. Could Derrida and Dennett be secret bedfellows with a shared dedication to the deconstruction of consciousness? Both play with self-touching arousals of self.

All living things have the power of auto-affection. And only a being capable of symbolizing, that is to say of auto-affecting, may let itself be affected by the other in general. Auto-affection is the condition of an experience in general. This possibility—another name for "life"—is a general structure articulated by the history of life, and provides a space for complex and hierarchical operations. (Derrida 1976, 165)

We can speculate that the greater virtues of sotto voce talking to oneself would be recognized, leading later to entirely silent talking to oneself. The silent process would maintain the loop of self-stimulation, but jettison the peripheral vocalization and audition portions of the process, which weren't contributing much. (Dennett 1991, 197)

My distaste for a terminological distinction between "adaptation" as original function and "exaptation" as supplemental function is akin to Derrida's rejection of originary meaning. Derrida wrote, "A meditation upon the trace should undoubtedly teach us that there is no origin, that is to say simple origin: that questions of origin carry with them a metaphysics of presence" (2016, 80). The search for original sense is the insistence upon a definitive answer to the question of the chicken or the egg. It is the demand for a univocal reading of a poem. It is the desire to assign blame and justify grievance in recursive recriminations of "he started it."

At a very late stage of composition of my book, I discovered I would not be the first, if ever there is a first, to detect a strong subtext of biology in Derrida. Francesco Vitale's (2018) *Biode-construction* is structured around a seminar taught by Derrida in 1975 on François Jacob's *Logic of Life*, but a close reading of Vitale's close reading of Derrida's close reading of Jacob's interpretation of life shall be deferred. Some readers will protest that I have misread Derrida, but I imagine him jumping to my defense. Texts have no meaning outside of interpretation. My text is a reading of Derrida's text being rewritten as you read. With due deference, "Jacques Derrida" is *not* (otherwise) *present* in this text, but his traces are everywhere.

(Genes (Memes (Memories (are inscriptions of the) personal) cultural) evolutional) past.

13   On the Origin of Meaning

There is grandeur in this view of life, with its several powers, having been originally breathed into a few forms or into one; and that,
whilst this planet has gone cycling on according to the fixed law of
gravity, from so simple a beginning endless forms most beautiful and
most wonderful have been, and are being, evolved.

—Charles Darwin (1859)

After the beginning but a very long time ago, the world contained RNA but neither DNA nor protein. By mindless processes
of natural selection, ineffective RNAs were degraded without
issue and effective RNAs acted in ways that, directly or indirectly, promoted their own replication. Each successful RNA's
self-promoting effects were its raison d'être, its function or purpose in life. Some RNAs acted as catalysts to facilitate beneficial
chemical reactions. Preferential copying of more efficient catalysts perpetuated RNAs that were able to discriminate *preferred*
substrates from *less useful* molecules, and perpetuated variants
that augmented their catalytic prowess by association with metal
ions or chemical cofactors. Other RNAs responded to things in
the world with choices of action. These choices were among the
earliest expressions of meaning, but, from so simple a beginning,

endless forms most beautiful and most hideous have been, and are being, evolved.

The primordial RNA world is long past, supplanted by more sophisticated forms of life, but direct descendants of some of its devices survive today in the "untranslated" regions of much longer messenger RNAs (mRNAs), where they control whether or not the "message" will be translated into protein. This chapter begins with a detour into biochemical minutiae to show how remarkably sophisticated feats of interpretation can be instantiated by allosteric macromolecules. Some readers may choose to skip ahead to subsequent sections that discuss the relation between the possible and the actual, and the importance of the arbitrary nature of the sign for the meanings of life.

Ribozymes and Riboswitches

An RNA that catalyzes a chemical reaction is called a *ribozyme*. Some ribozymes possess sequences that bind to specific small molecules with high specificity. These sequences are known as *aptamers* (from Latin *aptus*, "fitted": Ellington and Szostak 1990). An aptamer's binding partners are its ligands (from Latin *ligandus*, "fit to be bound"). The evolved fit of an aptamer for a ligand is a means whereby a ribozyme selects a thing from the environment for use in a chemical reaction. Once the ligand is bound, it participates in the chemical reaction. Aptamers are also used as sensors of things in the world that inform actions of an RNA without the ligand directly participating in a chemical reaction. Such an RNA, that functionally combines a sensor (aptamer) and an effector (expression platform), is called a *riboswitch* (Roth and Breaker 2009).

Ribozymes and riboswitches are molecular devices that cause a reaction if a ligand is present but not if the ligand is absent. A ribozyme is a *tool* used to effect an action, but a riboswitch is an *interpreter* that uses information in the choice of an action. The function of a ribozyme is to facilitate a reaction. The function of a riboswitch is to facilitate a reaction in the presence of ligand and to prevent the reaction in the absence of ligand. The distinction between a tool and interpreter is whether the "responses" to presence and absence of the ligand are both "intended," that is, whether nonoccurrence of the reaction in the absence of the ligand has conferred a fitness benefit. For the ribozyme, the absence of a ligand is an obstacle to achieving its function, but the riboswitch "prefers" that the reaction not occur if ligands are absent. A ribozyme acts in the world, whereas a riboswitch interprets the world.

Modern riboswitches reside in noncoding sequences of mRNAs and control whether the mRNA is translated as an enzyme (a catalytic protein). Glucosamine-6-phosphate (GlcN6P) is a substrate for the construction of bacterial cell walls. Its synthesis is catalyzed by the enzyme GlmS encoded by *glmS* mRNAs. The upstream-untranslated regions of *glmS* mRNAs contain an aptamer that binds GlcN6P. When bound to GlcN6P, the aptamer gains catalytic activity to cleave its own mRNA at an adjacent site promoting degradation of the mRNA and preventing translation of GlmS enzyme (Klein and Ferré-D'Amaré 2006). By this means, the *glmS* riboswitch implements negative feedback control. GlmS is translated if GlcN6P is absent but not if GlcN6P is present (Collins et al. 2007). (This device has features of both a ribozyme and a riboswitch because GlcN6P directly participates in cleavage of the mRNA but both the ON and OFF states are functional.)

In the simplest conceptualization of the *glmS* riboswitch, a change in conformation couples one bit of information about the world to one degree of freedom in action. Whether or not GlcN6P is present determines whether or not GlmS is produced. But the *glmS* riboswitch of *Bacillus subtilis* is more subtle than this. Its aptamer binds to either GlcN6P *or* glucose-6-phosphate (G6P) but inhibits translation only when bound to GlcN6P (Watson and Fedor 2011). The logic of this added complexity is that G6P is a substrate used to synthesize GlcN6P. Therefore, competition between G6P and GlcN6P for binding to *glmS* aptamers renders the population of riboswitches within a cell sensitive to the ratio of substrate to product. Translation of GlmS is switched off only when its product is abundant *and* its substrate is scarce.

B-group vitamins are chemically related to the ribonucleotides from which RNA is synthesized and are essential, as cofactors of protein enzymes, for metabolic processes shared by all living things. They probably functioned as cofactors of ribozymes in the RNA world (Monteverde et al. 2017; White 1976). A major class of riboswitches possesses aptamers for thiamin pyrophosphate (TPP), the biologically active form of thiamin (vitamin B_1). The aptamers exist as ensembles of rapidly interchanging states until binding of TPP stabilizes the "active" state. The transformation of the RNA energy landscape caused by binding of TPP results in a conformational change in the expression platform that mediates downstream effects on gene expression (Montange and Batey 2008; Winkler, Nahvi, and Breaker 2002). TPP riboswitches of prokaryotes directly regulate transcription or translation whereas those of eukaryotes regulate alternative splicing of mRNAs (Li and Breaker 2013; Wachter 2010; Wachter et al. 2007).

The thiM enzyme of *Escherichia coli* possesses binding sites for thiazole and adenosine triphosphate (ATP). This enzyme transfers a phosphate group from ATP to thiazole, creating thiazole phosphate from which TPP can be synthesized (Jurgensen, Begley, and Ealick 2009). *thiM* mRNA that encodes thiM contains a TPP aptamer that includes an anti-anti-Shine-Dalgarno (SD) sequence that is complementary to an anti-SD sequence of the expression platform that is complementary to an SD sequence immediately upstream of the translation start site of the mRNA. The SD sequence, when unbound to anti-SD, enables access of mRNA to the ribosome for translation into protein. Anti-anti-SD pairs with anti-SD in the absence of TPP allowing translation to proceed. Binding of TPP sequesters anti-anti-SD, allowing anti-SD to pair with SD, blocking translation (Winkler, Nahvi, and Breaker 2002). thiM is synthesized if TPP is sparse but not if TPP is abundant. But the double-negative logic—anti-SD (translation OFF) versus anti-anti-SD (translation ON)—hints that even this "simple" riboswitch performs more complex computations than use of one bit of information in binary choice.

Sometimes two riboswitches act in tandem. *metE* mRNAs of *Bacillus clausii* contain paired riboswitches for S-adenosyl methionine (SAM-e) and adenosylcobalamin (vitamin B_{12}). Either riboswitch terminates transcription when bound to its ligand. Transcription proceeds if neither SAM *nor* B_{12} is present (Sudarsan et al. 2006). The metabolic rationale appears to be that metE enzyme produces methionine, a precursor of SAM, but the bacterium possesses a more efficient pathway of producing methionine that uses B_{12} as a cofactor. Therefore, *metE* is transcribed and translated as metE only if the cell is deficient for both SAM *and* B_{12} (Breaker 2008).

Central to Ferdinand de Saussure's (1916, 1986) linguistics was the arbitrary nature of the sign, the lack of a necessary connection between signifier and signified. *Allostery* is a parallel concept in molecular biology in which binding at one site of a macromolecule causes functional change at another site (Goodey and Benkovic 2008; Monod and Jacob 1961). Allostery frees evolution by natural selection from stereochemical constraints because it enables physicochemically arbitrary associations of ligand and response. Jacques Monod, Jeanne-Pierre Changeux, and François Jacob's eloquent conclusion, presented for proteins but relevant to RNAs, is worth quoting at length:

A regulatory allosteric protein therefore is to be considered as a specialized product of selective engineering, allowing an indirect interaction, positive or negative, to take place between metabolites which otherwise would not or even could not interact in any way, thus eventually bringing a particular reaction under the control of a chemically foreign or indifferent compound. In this way it is possible to understand how, by selection of adequate allosteric protein structures, any physiologically useful controlling connection between any pathways in a cell or any tissues in an organism may have become established. . . . By using certain proteins not only as catalysts or transporters but as molecular receivers and transducers of chemical signals, freedom is gained from otherwise insuperable chemical constraints, allowing selection to develop and interconnect the immensely complex circuitry of living organisms. (1963, 324–325)

In the context of riboswitches, conformational coupling of aptamers to expression platforms allows physicochemically arbitrary coupling of signal to response. Allostery enables evolutionary mix-and-match between aptamers and expression platforms. GlcN6P is not physicochemically necessary for cleavage by glmS ribozymes because a mutated ribozyme that differs at only three nucleotides cleaves its mRNA in the absence of GlcN6P (Lau and

Ferré-D'Amaré 2013). Nor is there any physicochemical reason why glmS riboswitches could not reside in mRNAs that possess functions unrelated to the synthesis of GlcN6P. A glmS ribo-switch, for example, could be substituted for the TPP riboswitch of thiM mRNA, rendering the synthesis of thiamin contingent on the presence or absence of GlcN6P rather than TPP. But, neither a ribozyme that degraded glmS mRNA in the absence of GlcN6P, nor an mRNA that blocked production of thiamin in the presence of GlcN6P, makes adaptive "sense."

On an evolutionary timescale, the causal connections between informative inputs and meaningful outputs in the moment-to-moment functioning of biological interpreters are allosterically arbitrary but adaptively useful. Shifting shape makes sense.

On Hyperastronomic Numbers

Each increase in length from n to $n + 1$ nucleotides quadruples the number of possible RNA or DNA sequences (four nucleotides = 2 bits per nucleotide). Each increase in length from n to $n + 1$ amino acids increases the number of possible polypeptides twenty-fold (twenty amino acids $\approx$ 4.322 bits per amino acid). Sequences of n bits can convey $2n$ distinct messages. As a useful benchmark, the number of distinct strings of 300 bits approximates the number of elementary particles in the universe (Lloyd 2009). By this benchmark, the number of distinct 150-nucleotide RNAs and the number of distinct 70-amino acid proteins are of similar magnitude to the number of elementary particles. If you prefer a less-than-universal yardstick, a nonredundant pool of all possible RNA sequences 100 nucleotides in length would equal 1,013 times the mass of the Earth (Joyce 2002).

Most mRNAs are longer than 150 nucleotides, most proteins longer than 70 amino acids. An mRNA of the human *IGF1R* gene is more than 7,000 nucleotides in length and encodes a protein of more than 1,000 amino acids. The corresponding DNA sequence, including introns that are transcribed but spliced from the mature mRNA, is a staggering 316,000 nucleotides. The numbers of possible sequences of such lengths are hyperastronomic (Quine 1987, 224) or Vast (Dennett 1995, 109), but the human genome contains tens of thousands of protein-coding genes scattered as islands across uncharted oceans of noncoding sequence. Only a vanishingly small proportion of possible RNAs and possible proteins can have been explored within the extent and age of the knowable universe. Hyperastronomic spaces are "practically infinite," in the sense that there is no practical difference between search in a hyperastronomic or infinite space. Such spaces are never exhausted or encompassed.

Although numbers of possible RNAs, DNAs, or proteins undergo rapid ascents to hyperastronomic magnitudes as their length increases, hyperastronomic numbers are not the domain of evolution by natural selection. The number of sequences that have ever existed is undoubtedly very large, but still an earthly number. For a bit of genetic information to have been transmitted from the deep past it must have made a difference between "life" and "death" many times, first to spread through the population of sequences from its origin by mutation in a single sequence, and then to be maintained in the population against genetic drift and the onslaught of new mutations. The sequences that persist are products of historical processes that could have been different.

If there were only one way to build an aptamer for a particular ligand, then a 100-nucleotide aptamer (some aptamers are shorter, some longer) could never have been found by natural selection during the age of this planet, but it is not unusual to find arbitrarily chosen chemical activities in pools of the order of 10^{12} random RNA sequences of length 30–200 nucleotides (Knight and Yarus 2003). This means that there must be many possible aptamers with similar properties and that not every nucleotide is constrained by function. Even so, extant highly selective aptamers are unlikely to have been spontaneously generated in their current form. Rather, imperfect fits of earlier versions of aptamers for their ligands are likely to have undergone subsequent refinement by natural selection acting on "random" mutational variation in the "vicinity" of already functional sequences.

How can functional RNAs of thousands of nucleotides evolve? Most likely, natural selection first found functional sequences of much shorter length and then recombined these shorter sequences into longer sequences with more sophisticated functions (Lehman et al. 2011). The problem of selective search for novel functions among sequences of length $2n$ is much more manageable if it proceeds by recombination among already functional sequences of length n than by finding novel functions in a pool of random sequences of length $2n$. Examples of innovation by recombination include the coupling of old aptamers to new expression platforms and the formation of complex computational devices by concatenation of simpler riboswitches. Such evolutionary bricolage, using materials already at hand (Jacob 1977), will leave traces of a hierarchical modular organization in complex structures. By such processes, riboswitches have been combined evolutionarily and experimentally into more complex circuits to perform a wide range of digital and analog computations (Breaker 2012; Etzel and Mörl 2017).

The Potential and the Actual

The number of possible RNA sequences of a thousand nucleo-tides is hyperastronomically large, and each of these sequences can exist in a hyperastronomic number of possible three-dimensional conformations. The complete energy landscape of an RNA encompasses all *potential* conformations of its linear sequence in timeless space. Over infinite time, an RNA would occupy every point in its energy landscape an infinite number of times with the frequency of transitions between conforma-tions determined by the height of the energy barriers between them. But, in each infinitesimal moment of time, the RNA exists as some actual conformation. Life is not lived in infinite time and nothing happens in infinitesimal time. Between the infinite and the infinitesimal, the distinction between what *could be* and *what is* depends on timescale.

RNAs can be considered temporal ensembles of states, but the space of all possible conformations is much larger than could be actualized during the age of the universe for any RNA of appre-ciable length (Levinthal's paradox: Plotkin and Onuchic 2002; Zwanzig, Szabo, and Bagchi 1992). No RNA of significant length lasts long enough to realize its full potential, but real RNAs fold into kinetically accessible structures in reasonable time. Folding proceeds hierarchically by rapid formation of local secondary structures followed by slower tertiary interactions among the rapidly folded elements (Brion and Westhoff 1997). The energy landscapes of functional RNAs have forms that funnel folding from high-energy states to low-energy functional states by mul-tiple paths. These evolved mechanisms of self-directed assem-bly resolve Levinthal's paradox (Dill 1999; Leopold, Montal, Onuchic 1992).

Conformations separated by negligible energy barriers interchange in nanoseconds. Such ephemeral states are sampled in proportion to their probability density on a timescale of milliseconds. One might say that *actual* states at the nanosecond timescale exist as superpositions of *potential* states at the millisecond time scale. Both timescales are involved in RNA function. Nanosecond fluctuations enable the ribozyme to "find" its substrate and stabilize an unstable transition state to enable a chemical reaction to occur in milliseconds. The timescale of catalysis is longer than the timescale of substrate "recognition" because catalysis requires passage of a higher energy barrier and the thermal fluctuations that overcome such barriers occur less frequently. Other conformations are sheltered behind even higher energy barriers and are sampled more rarely, some at much longer timescales than the half-life of the RNA.

Energy landscapes shift in response to things in the world reshaping the accessibility or stability of alternative states. One might say that a molecule's experiences influence the realization of its potential. By such means, a ligand stabilizes one conformation of an aptamer out of an ensemble of rapidly interchanging conformations (Stoddard et al. 2010). At the timescale of the stabilized conformation, a *potential* structure is *actualized* by binding of the ligand. When a ligand stabilizes a conformation, the ligand *selects* the conformation (Csermely, Palotai, and Nussinov 2010). If the stabilized conformation has effects that contribute to the RNA being replicated, then the descendant sequences have been *selected* because of how the ancestral sequences responded to the ligand. In terms of mechanism, the ligand selects a conformation but, in terms of function, the sequence was *selected* because of how it responds to the ligand.

How Time Passes

> Time for you and time for me, and time yet for a hundred indecisions,
> and for a hundred visions and revisions, before the taking of a toast
> and tea.
>
> —T. S. Eliot

Different processes have different timescales. The selection of a
conformation of a riboswitch by a ligand, and the selection of
the riboswitch's sequence by how it responds to the ligand, illustrate a separation of timescales between conformational changes
in nanoseconds and evolutionary changes over hundreds to billions of years.

Human experience has a characteristic timescale. A fully
formed sensory experience develops in 100 to 200 milliseconds,
a single conscious moment lasts 2 to 3 seconds (Tononi 2004), a
life-span of three score years and ten clocks out at about 2 gigaseconds. By these measures, an elderly human has experienced
on the order of a billion conscious moments and, by retrieval
of stored memories, has conscious access to how things have
changed over a period of 2 gigaseconds. We can investigate processes that take place at longer timescales than a human life (such
as evolutionary changes in a riboswitch sequence) and shorter
timescales than a conscious moment (such as conformational
changes in the folding of a riboswitch), but these do not form
part of our phenomenal experience.

Mosses in my garden are fixtures for a millipede but, on the
timescale of growth, mosses explore space with complex behaviors of approach and avoidance. On this timescale of moss behavior, the movements of millipedes amid the moss are a potential

field of the environment, and, at the timescale of millipede meandering, molecular motions within millipede mitochondria are a potential field of the *milieu intérieur*. Most questions we care to ask about the world have a characteristic timescale. States that change more slowly can be treated as invariant. States that change more rapidly can be treated as a potential field of possibilities.

From the perspective of a separation of short-term from long-term changes, what is considered short term depends on personal predilections of timescale. Psychologists view individual behavior as short term and individual development as long term, whereas Bergstrom and Rosvall (2011) view individual development as short term and genetic transmission between the generations as long term. Within evolutionary biology, a separation of timescales exists between equilibration of gene frequencies on timescales for which the introduction of new mutations can be ignored (short-term evolution) and changes in phenotype over much longer periods for which change is driven by the flux of mutations (long-term evolution) (Eshel 1996; Hammerstein 1996). For a physiologist, short-term evolution is so slow as to be safely ignored in experiments, whereas for a paleontologist, short-term evolution is so fast as to be a potential field of evolving lineages.

Once upon a time, Saussure's (1916, 1986) school of structural linguistics distinguished a synchronic axis of simultaneity from a diachronic axis of succession. The synchronic axis concerned "relations between things which coexist, relations from which the passage of time is entirely excluded" (1986, 80). The diachronic axis concerned changes of language through time. "Everything is synchronic which relates to the static aspect of our science, and diachronic everything which concerns evolution" (1986, 81). For Saussure, the synchronic was an axis of ahistorical structure that existed outside of time, and the diachronic was the temporal axis of historical change.

Nothing happens outside of time. I have found two temporal interpretations of synchrony to be useful. In the simpler interpretation, synchrony refers to things that happen at faster timescales than diachronic change. On this interpretation, Saussure separates a synchronic axis of linguistic use in the here and now from a diachronic axis of linguistic change in the there and then. This interpretation parallels Mayr's (1961) separation of proximate and ultimate explanations in biology and Bergstrom and Rosvall's (2011) separation of a horizontal axis of development from a vertical axis of transmission. From this perspective, the causal connections between informative inputs and meaningful outputs of biological interpreters can be understood both as structural mechanisms, analogous to Mayr's proximate cause or Saussure's synchronic axis of simultaneity, or, as functional products of evolutionary history, analogous to Mayr's ultimate cause or Saussure's diachronic axis of succession.

In the second temporal interpretation of Saussure's atemporal axis of simultaneity, the synchronic encompasses things that change at timescales both faster and slower than the diachronic timescale. On this interpretation, diachronic change occurs in the interplay of sameness and difference, where the synchronic present comprises both the fixity of sameness (things that change at longer timescales) and potential fields of difference (things that change at shorter timescales). On this view, the diachronic drama of human history is enacted against a synchronic backdrop of things that change more slowly than the historical past and more rapidly than the existential present.

Decisive Action

A riboswitch's function depends on diachronic information from its evolutionary past and synchronic information from its environmental present. Evolutionary information is instantiated in the RNA sequence and represents the retrospective degrees of freedom of the replicative lineage, *what could have been*, and accounts for the riboswitch's repertoire of functional responses, *what could be*. Environmental information selects an actual response, *what is*, from these prospective degrees of freedom. From the synchronic perspective of mechanism, causal parity exists between the roles of ligand and aptamer in conformational change, but from the diachronic perspective of adaptive meaning, there is a *grammatical* separation of roles between the evolving RNA as *subject* and the unchanging ligand as *object*.

A riboswitch's energy landscape is its *form*. Some conformations have occurred sufficiently frequently to have been subject to natural selection, whereas other conformations have been sampled sufficiently infrequently to have been causally irrelevant for understanding the riboswitch's function. The evolved energy landscape, shaped by natural selection, takes advantage of aspects of form that are sensitive to small perturbations (butterfly effects) and aspects that are insensitive to large perturbations (bathtub effects). By these means a riboswitch can be insensitive to "irrelevant" perturbations but acutely responsive to "relevant" inputs. Its ligand flips the system from one basin of attraction to another.

Consider a paradigmatic riboswitch comprising an aptamer coupled to an expression platform. The aptamer undergoes reversible fluctuations (vacillations) among potential conformations until an actual conformation is stabilized by its ligand.

The ligand-induced stabilization of the aptamer causes an allosteric shift in the expression platform that causes an irreversible chemical reaction such as termination of transcription. The evolved structures of aptamer and expression platform are respectively *uncertain* and *undecided* until detection of the ligand by the aptamer causes an irreversible reaction mediated by the expression platform. The riboswitch has instantiated a reasoned choice in which a *decision* (irreversible action) occurs for a *reason* (detection of the ligand). The allosteric mechanism that couples input to output, that converts vacillation to decisive action, is physicochemically arbitrary but makes adaptive sense.

Let us concede to advocates of reductionist mechanism that everything that exists in the world can be considered an outcome of nothing but efficient and material causes. Structural biologists can explain the mechanism of a TPP riboswitch at an atomic level. They can describe how binding of the ligand stabilizes a transient conformation of the aptamer that is selected from among an ensemble of rapidly interchanging possible conformations. But this explanation of synchronic mechanism (*how* it works) leaves unanswered the historical *how come?* and the teleological *what for?*

Extant TPP aptamers are not products of recent spontaneous generation. Rather, an RNA sequence that bound TPP was discovered four billion years ago, give or take a few million years, and has been the progenitor of all existing TPP aptamers. In this four-billion-year history there has been both a continuity of genetic transmission and a continuity of TPP binding to aptamers. A full account of *how come?* in terms of efficient and material causes is unattainable, and the details, if such an account were possible, would be of little consequence. Such an account would need to keep track not only of the survival and reproduction of all the ancestors of current aptamers but also of the fate of mutant aptamers that left no descendants in the struggle for existence because, in an ecologically constrained world, the demise of the losers is a necessary part of a causal account of the success of the winners. TPP riboswitches enhance metabolic efficiency because they shut down synthesis of thiamine when thiamine is not needed. The proximate causes of death of organisms that lack this ability will have been many and varied, resulting from idiosyncratic combinations of circumstances in which slight differences in metabolic efficiency made the difference between life and death. The difference that made the difference in all these life-or-death outcomes was the difference between binding and not binding to TPP.

All TPP aptamers are descendants of an ancestral aptamer that evolved in the RNA world before the origin of DNA and proteins. These highly conserved structures are now associated with diverse expression platforms that regulate thiamin metabolism in bacteria, archaea, and eukaryotes (Duesterberg et al. 2015; Winkler, Nahvi, and Breaker 2002). An RNA sequence that recognized TPP was discovered by selective search more than two billion years ago and its descendants have persisted ever since, despite genetic drift and the constant introduction of nonfunctional variants by mutation. The evolutionary *maintenance* of TPP aptamers is explained by their affordances, the aptamers' aptnesses, in particular the useful handle an aptamer provides for functional engagement with TPP. I will make the deliberately provocative claim that everything other than this affordance is causally irrelevant to understanding why TPP aptamers remain billions of years after their origin. TPP aptamers exist and persist for the sake of binding TPP.

Monkeys and Typewriters

> Numerous and varied are the objections that have been advanced
> against the theory of selection . . . to the opposition of our own
> day, which contends that selection cannot create but only reject, and
> which fails to see that precisely through this rejection its creative
> efficacy is asserted.
> —August Weismann (1896)

Darwinism has often been criticized as claiming that things of
value can be produced by purely random processes. The Princeton theologian Charles Hodge (1878, 52) rejected Darwin's reli-
ance on the "gradual accumulation of unintended variations of
structure and instinct":

In like manner we may suppose a man to sit down to account for the
origin and contents of the Bible, assuming as his "working hypothesis,"
that it is not the product of mind either human or divine, but that it
was made by a type-setting machine worked by steam, and picking out
type hap-hazard. In this way in a thousand years one sentence might
be produced, in another thousand a second, and in ten thousand more,
the two might get together in the right position. Thus in the course
of 'millions of years' the Bible might have been produced, with all its
historical details, all its elevated truths, all its devout and sublime poetry, and above all with the delineation of the character of Christ, the
ιδεα των ιδεων [idea of ideas], the ideal of majesty and loveliness, before
which the whole world, believing and unbelieving, perforce bows down
in reverence. And when reason has sufficiently subdued the imagination
to admit all this, then by the same theory we may account for all the
books in all languages in all the libraries in the world. Thus we should
have Darwinism applied in the sphere of literature. This is the theory
which we are told is to sweep away Christianity and the Church! (Hodge
1871, 61)

Hodge, and many similar critics of natural selection, interpreted
Darwin as assigning creativity to randomness. They could not

see how a mindless process could generate order from randomness. Only a mind could be creative.

Hodge's type-setting machine could have been unproductively employed in the printing press of Jorge Luis Borges's (2000) Total Library, the repository of all possible books mechanically produced by randomly stringing together letters and punctuation marks. Somewhere on the shelves of that library would exist every book that has been written, every book that will be written, and every book that could be written. It would contain a copy of this book and every one of its rejected drafts. Somewhere on its shelves, tantalizingly out of reach, would be a version of this book that would convince all its readers of my intended meaning. But the library is completely useless.

Everything would be in its blind volumes . . . but for every sensible line or accurate fact there would be millions of meaningless cacophonies, verbal farragoes, and babblings. Everything: but all the generations of mankind could pass before the dizzying shelves . . . ever reward them with a tolerable page. (Borges 2000, 216)

For all practical purposes, nothing makes sense in a Total Library. Some principle of selection is required to find value in randomly generated texts.

A different reason, perhaps the same reason in different guise, for rejecting natural selection as creative is to view selection as a purely negative process that merely eliminates variation generated by other means that are the true source of creativity. A sample of quotations will give the flavor:

The function of natural selection is selection and not creation. It has nothing to do with the formation of new variation. It merely decides whether it is to survive or be eliminated. (Punnett 1913, 143)

[Natural selection] is essentially a negative substitute for teleology: it accounts for the disappearance only and not for the emergence of forms—it suppresses and does not create. (Jonas 1966, 51)

My problem is that some Darwinists are inclined to attribute creative powers to what they call natural selection. . . . The only creative element in evolution is the activity of living organisms. (Popper 1986, 119)

Where does adaptive change come from? A trivial but sometimes overlooked point is that it never comes from natural selection . . . natural selection cannot create anything. (Dupré 2017, 5)

Most of these critics ascribe innovation to the mutational processes that generate the variation that is uncreatively accepted or rejected by natural selection. They implicitly ascribe creativity to the authors of the books in Borges's Total Library. The view that mutation is the source of meaning misunderstands meaning. Mutation is nonmeaning. At the very beginning, in the origin of difference, is nonsense.

Semantic Topiary

[Natural selection] is far more "creative" than the pruning of a tree, to which it has sometimes been compared, and more creative even than the whittling of wood from a block to form an image which, among an infinite number of other potential images, had, in a sense, lain latent within the block. If this is not actual creation, then no sculptor creates his statues, and no poet, in selecting his particular words out of an almost infinite number of possible combinations creates his verses.

—Hermann Muller (1949)

Consider the token tree of all existing TPP aptamers and trace its branches back, by converging paths, to their last common RNA ancestor, the *urtoken*. The immediate predecessor of an extant aptamer will be a DNA sequence; then this DNA sequence will have exclusively DNA ancestors, until one gets back to a world of exclusively RNA ancestors. The RNA urtoken undoubtedly already had a high affinity for TPP, but as we trace its ancestry further back in time we would eventually come to sequences with less and less affinity for TPP, until we finally came to a stem-token with no appreciable affinity (figure 13.1).

Now start with this stem-token and follow its descendants forward in time. The branches of this tree will have been heavily pruned by natural selection. Most mutations would have little effect on affinity for TPP or would reduce affinity, but mutations that reduced affinity would preferentially be found on branches that were cut out of the tree. There will be one unique path that leads forward from the stem-token to the urtoken and that replays forward the backward "tape of ancestry." The rare mutations that increased affinity for TPP would preferentially be found on this path. If one had pruned the token tree at random, ever so many times, one would never have obtained the sequence of the urtoken by mutation because of the hyperastronomic size of the sequence space. But the trick is turned when the tree is pruned by the environment, because new random mutations now occur on branches that had already been nonrandomly selected because they possessed mutations for greater aptness.

The mutations that occur on the path from stem-token to urtoken exibit a trend toward increasing affinity for TPP. How can one account for *directional mutation* toward greater aptness? The answer is simple. Mutation is a locally random process with respect to aptness, but the series of mutations found on successful paths proceed haltingly in the direction of greater aptness. The directionality comes from the environment that selects, not from the mutational process that generates branches of varying aptness. Hermann Muller, who was to receive the 1946 Nobel Prize in Physiology or Medicine for his discovery of X-ray

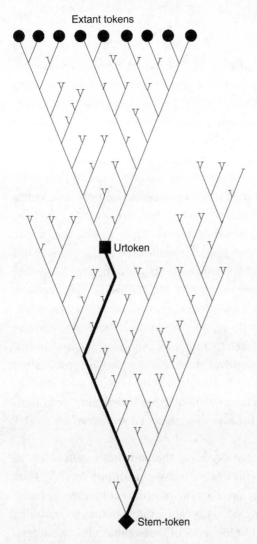

Figure 13.1
Simplified token-tree of TPP aptamers. Extant TPP aptamers (tokens) are represented by filled circles at the top of the figure. Their ancestry is then traced back to their most recent common ancestor, the urtoken (filled square). The ancestry of the urtoken is traced back to a stem-token without affinity for TPP; then the descendants of the stem-token are traced forward in time. One forward lineage (the stem lineage) connects stem-token to urtoken. This lineage contains a series of mutations that *progressively* increase affinity for TPP.

mutagenesis, expressed the implications this way: "[It is] the peculiar power of multiplication of mutant forms which turns this trick of converting accident into order, by making such very extraordinary combinations of accidents possible as could not otherwise occur" (1929, 498).

Many considerations come into play when pruning a tree. The aptamer was shaped by what was rejected as much as by what was retained. The selective process eliminated unwanted associations with similar ligands that did not work as well or that spoiled the effect. In the evolved fit of a TPP aptamer for its ligand, many parts must work together to conform to the form of the ligand. The aptamer's text has been subject to selection on all of its disparate effects. Some parts must do triple duty in molding the form to the ligand, in coupling the aptamer to the expression platform, and in interacting with other aptamers. Natural selection is a poet who tries the mutations in search of a *bon mot*. Riboswitches, genes, and organisms are the poetry of life. They mean many things at once. Perhaps this is the ultimate cause of disagreement within evolutionary theory. There are many ways to interpret a poem.

The Creativity of Natural Selection

The power of Selection, whether exercised by man, or brought into play under nature through the struggle for existence and the consequent survival of the fittest, absolutely depends on the variability of organic beings. Without variability nothing can be effected.
—Charles Darwin (1883)

The origin of meaning can be ascribed to natural selection sorting meaningful from meaningless mutations. Differential copying preserves variants of value and gives directedness to the sequence of mutations in evolutionarily successful lineages. This is a process that separates gold from dross. Muller addressed the standard argument from improbability head on:

In beings without the property of multiplication of variations, and its corollary, natural selection, any such incredible combination of accidents as ourselves would have been totally impossible of occurrence within the limits of practically any number of universes. We are thus really justified in feeling that we could not have fallen together by any accident of inanimate nature. But, given the power of multiplication of variations resident in "living" things, due to their genes, and all this is changed, and we are enabled to enjoy the benefits—such as they may be—of being the select of the select, such as it would have taken a surpassingly vast number of worlds to search through, before our match could be found anywhere by the ordinary processes of chance. (1929, 504)

Ronald Fisher (1934) similarly criticized theories that ascribed the "effective guidance of the evolutionary process to the agencies which cause mutation." He acknowledged mutation "as a condition which renders evolution possible" but, when he came to locate "in time and place the creative causation to which effects of especial importance are to be ascribed" he found it to be "in the interaction of organism and environment—in the myriad biographies of living things—that the effective causes of evolutionary change must be located." Systems of mutable replication exhibit "spontaneous creativity" in their selected responses to the environment. Fisher later returned to this theme:

Just where does the theory of natural selection place the creative causes which shape evolutionary change? In the actual life of living things; in their contacts and conflicts with their environments, with the outer world as it is to them; in their unconscious efforts to grow, or their more conscious efforts to move. Especially in the vital drama of the success or failure of each of their enterprises. (1950, 17)

He further wrote: "Living things themselves are the chief architects of the Creative activity" via their "willing and striving" (intentions) and their "doing or dying" (actions). "It is not the mere will but its actual sequel in the real world, its success or failure that is alone effective" (19). It is the engagement of living things with their environment, mediated by the very real consequences of death or survival, that has shaped living things.

> Weep not that the world changes—did it keep
> A stable, changeless state, 'twere cause indeed to weep.
> —from "Mutation" by William Cullen Bryant (1794–1878)

14   On the Past and Future of Freedom

I was born by the river in a little tent,
And just like the river I've been running ever since
It's been a long, long time coming,
But I know a change gonna come, oh yes it will.

—Sam Cooke (1931–1964)

Life is interpretation. From a beginning in the RNA world, an interpretative arms race unfolded among increasingly complex, heritable agents that strove to *make sense* of the previously uninterpretable. Agents that were informed by new inputs, or interpreted old inputs in new ways, could exploit resources "invisible" to less perceptive agents or avoid dangers to which the imperceptive succumbed. In the process, RNAs that at first instantiated both text (preservation of information) and performance (action in the world) were relegated to roles as messengers between DNA (as the archival record of past natural selection) and proteins (as effective actors). The complexity and precision of interpretation was facilitated by the evolution of high-fidelity interpreters of genetic texts: ribosomes for translation of mRNA into protein, RNA polymerases for transcription of DNA into mRNA, and DNA polymerases for copying the archival text. The replacement of RNAs by proteins marked a major expansion of the chemical lexicon from the four ribonucleotides of RNA to the twenty amino acids of proteins. Not only were there $20n$ possible peptides of length n, compared to $4n$ RNAs of the same length, but the twenty different side-chains markedly increased the expressivity of the chemical language. Twenty amino acids could do more different things than four ribonucleotides.

Complex interpreters were cobbled together from simpler interpreters by evolutionary bricolage and subsequent refinement. Organisms were not simply passive consumers of information presented by the environment but inquisitive seekers of tidbits that might be useful. Sophisticated interpretations of sensory input as behavioral output depended on simpler interpretations by suborganismal interpreters. The outputs of one subsystem became inputs to other subsystems: allosteric proteins and RNAs collaborated in regulatory networks to interpret the binding of ligands to cell-surface receptors as changes in gene expression within cells; the release of neurotransmitters and neuromodulators at many synapses were integrated as decisions of whether or not a neuron fired; groups of neurons interacted to make higher-level decisions by complex processes of feedforward and feedback; molecular memories recorded past interpretations for future use as input to decision making.

Each bit of genetic information that has persisted from the deep past required many selective deaths to reach high frequency in the gene pool and continuing selective deaths to be maintained in the gene pool. The evolution of individual learning provided a major advance in interpretative sophistication. Selective deaths were no longer necessary for selection-by-consequence to refine the inner workings of interpreters. Genetic information from the evolutionary past and memories of the personal past could now both be brought to bear on the interpretation in meaningful action of data of immediate experience. Learning from others expanded the sources of useable information to include cultural traditions. Memory aids response to that which persists despite change or that which recurs after change. When we encounter the never-before-encountered, we search for metaphors in what we already know.

Each species is unique, but something is special about humans. Culture and language lie at the heart of this special something. Hints of these are present in other species, but we have crossed a threshold and become something new under the sun. The expressivity of the flux of meaning exploded with human language. Each word has distinctive uses. A language with a limited lexicon of 10^4 words can generate 10^{4n} strings of length n. Most strings are ungrammatical, just as most random strings of nucleotides or amino acids do nothing useful, but the number of possible meaningful strings has already escalated to hyperastronomic numbers for strings the length of this sentence. Spoken texts were mostly ephemeral, although some oral communities sustained bards who memorized rhymed texts of great length. Writing was a major advance because it allowed long-term storage of strings of indefinite length on external media. We became people of the book (and now of the internet).

When called before the Diet of Worms in 1521, Martin Luther is said to have said "Hier stehe ich, ich kann nicht anders." (Here I stand, I cannot do otherwise.) Was his *determined stance* a free act? Luther believed he acted from necessity. His will was bound by God's will. Many materialists would substitute physical law for divine fiat and agree that Luther could not act freely because his actions had prior causes. Another interpretation is that we are free when our actions are not controlled by immediate causes

external to ourselves—*when we act out of who we are* for our own
ends. By this interpretation, Luther's refusal to recant expressed
his freedom from external control. His causes determined his
actions. The assembled forces of the Holy Roman Empire and
Catholic Church could not bend him to their will.

Our formal causes, the textual record of our genetic and personal narratives, are a past source of present action and future
intent. When we act in response to immediate stimuli, what we
choose expresses who we are, and the efficient causes of who
we are are not manipulable by others here and now because
those causes were there and then. Our formal and final causes,
as recorded in our genetic material, are part of who we are; but so
too are the *formative* experiences of our individual lives. Things
that happened fifty years ago and a billion years ago inform
my choices here and now. Things that happened long ago are
beyond your control, but they are also beyond my control. Have
I simply replaced external control by proximate causes with
external control by remote causes? To think about this question
requires close attention to explanatory timescales.

Consider the long period, reaching back to the beginnings of
life, over which changes in our genetic makeup occurred that
are shared by all members of our species. This is a timespan over
which information *from the environment* was incorporated *into
our genes*. Some changes that happened near the beginning of life
are shared by most living things. We share more recent changes
with slugs and even more recent changes with chimpanzees. We
do not need to go back very far before nothing that happened
accounts for any *differences* among us, because we all have the
same ancestors who experienced the *same environments*. What we
inherit unchanged from these ancestors is part of the bedrock of
who we are as human beings and beyond our control. It is the

formal cause of our being human, a cause of our sameness. We call it *human nature.* The environmental factors responsible for the selective maintenance of human nature are the final causes of our being human.

Our formal genetic causes (informational genes) are the archival texts of past natural selection and are instantiated as genetic causes (material genes) during our individual development. These material genes specify the construction of real-time interpreters, ourselves, who strive to survive in an unpredictable world. Our material genes participate as efficient causes in our self-construction but rely on dependable elements of the environment as supplementary efficient causes of our developmental sameness. Each interpreter needs to be flexible in decision making because the environment it will encounter cannot be anticipated in detail. Its structure must be modifiable by experience of what worked and did not work in the recent past. Because we are human interpreters, we are constructed so as to learn from others and modify our structure and behavior using culturally communicated information, much of it transmitted by language.

What are the difference-makers during development that account for our *individual natures*? Some of these are genetic differences that arose more recently than the timescale of human nature and some are environmental differences that occurred more recently than the timescale of our genetic differences. We often think of our childhoods as *happening to us* and *beyond our control* because we were not fully formed. Much of the controversy about the causal role of genes in development involves people talking past each other because they have different concepts of cause and assume different explanatory timescales. Some are concerned with causation as mechanism and others with causation as difference making. Some are concerned with human nature and the causes of developmental sameness, whereas others are concerned with individual natures and the causes of developmental difference.

A very good friend of mine has a *de novo* mutation of one of his *CHD7* genes, probably the copy he inherited from his father (*de novo* means the mutation occurred within the last generation). This mutation causes impairments of visual, auditory, olfactory, vestibular, and proprioceptive inputs. It has made a major difference in the information he receives from the world and the actions he can take in the world. He was dealt a very bad card but has played a very good hand (well done!). Why did such a tiny change to a material cause make such a major difference in development? The mutation perturbed an ancient formal cause and profoundly disrupted the interpretation of how to develop by cells somewhen in his developing nervous system. Its effects have no final cause. There is no selective history of maintaining this mutation. It conveys no useful information about the past.

Environmental differences acting as efficient causes can have profound effects on development. Some children raised in Romanian orphanages under Nicolae Ceaucescu had their lives disrupted more severely than my son with the *CHD7* mutation. (Liam tells me "it was not such a bad card.") Thankfully, tiny environmental differences, even substantial differences, do not have large developmental effects. The environment must be severely perturbed to make such a large difference. Human nature protects us from most minor insults. Children born in Nepal—a difference, not an insult—grow up into "typical American kids" in the United States. Their Nepali heritage and their Nepali ancestry will account for some differences and their American upbringing and human nature for the samenesses.

At the timescale of moment-by-moment behavior, genes have little control. We make many decisions without consulting our genes. Smiles are exchanged between a mother and child without genetic input. Material genes, as efficient causes, are not actors on the stage. They were tools used in the construction of the stage-sets and can be used as tools to remodel the sets for future performance. We, as the real-time interpreters of our world, are the performers in the play. But this is an improvised drama without a written script. I cannot decide what to say until I have heard what you say. Causal stories can be told in many ways.

The soul is the form of the living body. The word "soul" has accreted associations from two millennia of Christian theology that are not part of my intended meaning, which is closer to Aristotle's *psuche*, the word that was translated as *anima* in Latin and *soul* in English. *Psuche* (from which we derive "psychology") was the breath of life, the source of animation, that which distinguished a living body from a dead body. *Psuche* initiated the activity of living bodies; it was the essence that made of the body one kind of thing rather than another; and it was the end "for the sake of which" the body acted. This *telos* could be interpreted as either the beneficiary of the action or the utilitarian purpose for which the action was performed. *Psuche* was the efficient, formal, and final cause of which *soma* was the material cause and the union of *soma* and *psuche* was the living thing (Aristotle, *On the Soul*). Plants also had souls. Souls are the intricate organizations of living things that inform their choices whether conscious or unconscious. The fully formed soul develops during ontogeny and can deteriorate with age before finally ceasing to be.

Form is simply organization of matter, soul simply a particular level of integration of form. In the more complex forms of life, the gain of soul is gradual, and the loss of soul may also be gradual. Society must grapple with different beliefs about human ensoulment and different definitions of death. Form persists in recently dead bodies—parts of which are still living and actively maintaining their form—but the bodily integration has been lost. One might say that cellular souls can survive the death of the bodily soul. Some of the still living parts can be transplanted to become parts of other bodies and thus sustain other bodily souls.

When we ask whether we are free, we are usually asking whether our soul is in control of our day-to-day choices. In what senses are the choices of souls determined by external efficient causes? At an abstract level, our souls make use of bathtub effects as cancelers of difference and butterfly effects as amplifiers of difference. Bathtub effects allow souls to be unperturbed by unwanted causes. They buffer our souls from the slings and arrows of outrageous fortune. It makes no difference how water enters a tub; it leaves by the drain. Butterfly effects render us sensitive to relevant information. The mere flap of a wing, perhaps the movement of a few ions across a membrane, makes a large difference. They are decision points that use minimal means to shift from one tub to another. Butterfly and bathtub effects are the mechanisms of the soul that determine how we respond to the world. Whether we respond, and the way we respond, is determined by the structure of our souls. If our past had been different, we would respond differently. We would have different souls.

Genes are specifications of souls. They provide information for building an interpreter that decides for itself. The choices to

be made are not part of the specifications, because the configuration of future events cannot be anticipated. What is provided
are tools used in choosing. The specifications include modifi-
ability by experience (but not just any experience). The structure
of the soul is jointly determined by an evolutionary history that
has written the genetic specifications and a developmental history that has revised the specifications and remodeled the soul
in response to experience. And these histories have produced
souls that are partially freed from history. Souls have an imma-
nent rather than external *telos*. Souls make of their lives what
they will.

We are not slaves to our genes, because they have delegated
decision making to our souls. We are not slaves to our culture,
because our souls judge and decide which parts of culture to
accept and which to reject. Our genetic specifications allow
modification by culture because souls that ignored cultural
norms were less successful than those that selectively conformed. But our modification by culture is selective, because
souls that too slavishly followed the dictates of culture were less
successful than those who made up their own minds and, by
so doing, sometimes transformed the culture. We are cultural
conformists and cultural skeptics by *human nature*. We are freed
from genetic determinism by culture and individual choice. We
are freed from cultural determinism by human nature. This is a
division of powers. Our souls are free within constraints, chief
among which are the freedoms of other souls.

Your soul (who you are) is an unmoved mover (here and
now). You are synchronically free to interpret information as
meaningful choice because your soul is diachronically empow-
ered to make sense of an unanticipated world as an expression
of your individual nature and our shared human nature. Are you

enchained because you cannot be other than human or other than yourself?

Moral codes are important cultural inputs to the remodeling of souls. These are human inventions that are part of the evolving technology of the soul. More than one invention can serve similar ends. To recognize moral codes as inventions is not to belittle them. Inventions have made a profound difference in how our lives are lived. But this is to say that moral codes are associated with trade-offs, design flaws, fads and fashions, like any other technology. Technologies have a tendency to become increasingly sophisticated over time. Moral codes are subject to similar "progress." Moral "absolutes" occur in the synchronic here and now, grounded in the diachronic relativism of cultural and evolutionary change.

There is no *practical problem* of evil. We know perfectly well what to do about evil. We must recognise it, in ourselves and in the world and repudiate it. To the best of our ability we must combat it, study it, and extirpate it. That is exactly what we mean by the word Evil: that which we are called upon unequivocally to attack and eliminate. On this view evil is relative, it changes its nature with evolutionary progress and with the changing structure of human society. Attempts at codification such as the Ten Commandments, or the Seven Deadly Sins, may remain valid for a long while; manifestly we cannot expect them to be adequate for ever. (Fisher 1950, 21–22)

We cannot do just anything. Our choices are limited by the capabilities, and the resistances, of our souls. How should we undertake to understand the workings of souls? The standard scientific approach has been to obtain ever-more detailed information about a soul's material parts. We have learned much by these methods, but it is no coincidence that technologies which have sought to understand the relations among inputs and outputs of our souls, treated as black boxes, have been more effective at fine-tuning our behavior than those that tweak the mechanisms. It is hard to understand how and why complex nonlinear systems behave as they do.

Music changes behavior because it speaks directly to our souls rather than attempts to swat a butterfly or remodel a bathtub. Rhythm, melody, and harmony are not in the parts but the relations of parts. Souls orchestrate the symphonic counterpoint of living beings. They improvise and extemporize on standards from the past. From whence comes the organ in organism? Greek *organon* was a tool or an instrument. It was "that with which one worked," derived by ablaut from *ergos* for "work." Latin *organum* could be a mechanical device, an engine of war, or a musical instrument. Old English *organ* was a musical instrument, and also a melody or song, but its definitions were extended in the fifteenth century to parts of a body that had an instrumental function (hence *organic*). The verb *to organize* was "to furnish with organs," from which was derived *organization* and *organism*. From whence comes the text in context? The primary meaning of context in the *Oxford English Dictionary* is the "weaving together of words and sentences; construction of speech, literary composition. *Obs.*" Cognate terms include textile and texture. Their etymological connection is through weaving.

All human beings, indeed all organisms, are constantly processing untold amounts of information in order to exist and function in the world. Our interpretations of inputs as outputs are internal processes conditioned by our unique evolutionary and experiential histories. I interpret the world as I do because of who I am and would respond differently if I were other than who I am. If I am to understand how and why you interpret the world as you do, I must understand who you are. Our entropies of observation and action have sufficient degrees of freedom to ensure that the number of things to which we could respond and the number of ways we could respond are both hyperastronomic numbers. It would be senseless to complain that our freedom of action is constrained because the space of our options is hyperastronomic rather than infinite. I possess enough degrees of freedom to get on with my life.

Freedom from control is freedom from being used as a tool of others' intentions. We have evolved to decide for ourselves, not to be easily manipulated. But we are beset by efforts of others to control or influence our choices, whether this be by crude coercion, deception, indoctrination, or persuasion. The sciences of human nature reveal foibles of our mechanisms, simple ways to subvert our choices, and these sciences inform modern technologies of marketing, drugs, and political spin that exploit vulnerabilities of our mechanisms to nudge our actions for our good or our ill. We are treated as tools rather than free agents. Some of these technologies regard our inner complexities as black boxes and exploit statistical regularities between inputs and outputs. Other technologies attempt to peek inside the box to understand its mechanisms and tweak its outputs. All these technologies of control can be predicted to become progressively more sophisticated because some people will pay good money to change how we choose. Must we surrender our freedom and become mere means for the ends of others? Can we save our souls from being hacked? The narrative arc of human history leaves us hanging on the brink as the credits roll.

15 Darwinian Hermeneutics

> Teleology is a lady without whom no biologist can live. Yet he is ashamed to show himself with her in public.
>
> —Ernst Theodor von Brücke (quoted in W. B. Cannon 1945)

Francis Bacon (1605) disparaged Minerva and the Muses as barren virgins and thus relegated the arts to the category of things irrelevant to physical inquiry. René Descartes's (1641) separation of the mind as thinking thing from bodily mechanism had similar implications. Creativity belonged to the ghost, not the machine. And thus, near the beginnings of the Scientific Revolution, we find early intimations of an estrangement between humanistic and scientific approaches to knowledge.

The liberal arts (*ars liberalis*) of the medieval university combined the *trivium* (grammar, dialectic, rhetoric) and *quadrivium* (arithmetic, geometry, music, astronomy). The fifteenth and sixteenth centuries saw reform of the *trivium* as the *studia humanitatis*, forerunners of the modern humanities. This curricular reform involved a shift from emphasis on logical disputation (dialectic) to the reading and interpretation of classical texts. Poetry, history, and moral philosophy joined the educational attainments befitting a free man (Kristeller 1978; Nauert 1990). The seventeenth century saw major developments in natural philosophy, the precursor of the modern sciences. The social sciences claimed a place at the academic table during the nineteenth century. These divisions of scholarship continue to jostle for appointments in the academy and coverage in the curriculum.

German universities of the nineteenth century were the arena of extended polemics between advocates of the *Naturwissenschaften* (natural sciences) and *Geisteswissenschaften* (roughly the humanities and social sciences; *Geist* can be translated as "ghost," "spirit," or "mind"). I am unable to fully *understand* these debates and adequately *explain* them to you, because of my very limited abilities in the German language. In my attempts to interpret a long German sentence, I can tease out possible meanings of some of the phrases but am unable to identify the intended meaning because the meaning of each *part* relates to the sentence as a *whole*, and the sense of the sentence must be understood as *part* of an extended argument, the thesis as a *whole*, that contains many sentences much longer and more complex than the sentence you have just read. English translations, when available, break up the long sentences into parsable parts but at the cost of another layer of interpretation interposed between my reading and the pretranslated text.

Hermeneutics developed as a methodology for the study and interpretation of written texts, especially sacred scriptures and the writings of classical antiquity. Its central problematic was how to make sense of a text written long ago in an alien language. Interpretation was envisaged as a recursive process in which the sense of individual words and phrases was gained from the sense of the whole but the sense of the whole was constructed from the sense of the individual parts, all in the context of reading other texts. The reciprocal relation between interpretation of parts in the context of the whole and the whole in terms of its parts came to be known as the *hermeneutical circle*. The scope of hermeneutics expanded to include the interpretation of all social phenomena, not just written texts. This expansion in the domain of hermeneutics has been accompanied by expanded definitions of text. At the limit, anything that is interpreted is a text, including the course of human history and all human actions, whether intentional or unintentional (Ricoeur 1971). My personal definition of a text—an interpretation intended to be interpreted—is narrower because it excludes the unintended.

Wilhelm Dilthey wrote extensively on the distinctive methodology of the *Geisteswissenschaften* and related these methods to general problems of hermeneutics. His statement "Die Natur erklären wir, das Seelenleben verstehen wir" (Dilthey 1894) is the *locus classicus* of an influential distinction between *Erklären* (explanation) as the explanatory mode of the *Naturwissenschaften* and *Verstehen* (understanding) as the elucidatory principle of the *Geisteswissenschaften*. In English translation, with succeeding sentences:

We explain nature but we understand mental life. Inner experience grasps the processes by which we accomplish something as well as the combination of individual functions of mental life as a whole. The experience of the whole context comes first; only later do we distinguish its individual parts. This means that the methods of studying mental life, history and society differ greatly from those used to acquire knowledge of nature. (Dilthey 1979, 89)

In Dilthey's view, explanation was synthesis, a building-up from disconnected parts, whereas understanding began from a connected unity. Understanding was analysis of the parts of this singular whole. In "Die Enstehung der Hermeneutik" (Dilthey 1900; translated as the "The Rise of Hermeneutics," 1996), he formulated the central problem of the *Geisteswissenschaften* (human studies) as generalization from singular existence:

When the systematic human studies go on to derive more general lawful relations and more inclusive connections from this objective apprehension of what is singular, the processes of understanding and interpretation still remain basic. Thus, these disciplines, like history itself, depend for their methodological certainty upon whether the understanding of what is singular may be raised to the level of universal validity. So at the threshold of human studies we encounter a problem specific to them alone and quite distinct from all conceptual knowledge of nature. (235)

But this problem is not unique to human studies. It arises for all living things. Each species, each gene, is an individual with a deep evolutionary history. Each organism is an individual with a unique developmental history. A biologist confronted by the behavior of a slug is in much the same position as a drama critic. The individual performance was shaped by evolutionary and developmental pasts, unknowable in detail and unmanipulable by the methods of experimental science. Knowledge from diverse sources must be brought to bear on problems of interpretation if one is to understand the meanings of a slug. *Biologie* is a fruitful scion of *Geist* grafted on *Natur*.

The hermeneutic circle—the reciprocal relation between understanding the parts from the whole and the whole from the parts—is central to understanding life. The metaphor of natural selection subsumes all processes by which organismal performance determines which organisms survive to reproduce and hence which genes are copied and recopied. The whole selects the parts. The problem selects the solution. Informational genes are the archival text of past performances that *formed* the text that informs. Material genes are actors in the play. Life is a cycle in which text and performance are reciprocally cause and effect of each other. The circle is rescued from eternal recurrence of the same by mutation (origin of difference) and selection (generation of meaning by the erasure of difference). The stage props that have withstood the tests of repeated use are tools that organisms use to interpret their world.

The intricate mechanisms of living beings, what I have called souls, enable organisms to integrate sundry sensory inputs as choices of unified action. Souls are not easily analyzed: they are explicable and inexplicable in purely physical terms. Soul-structures are physicochemically arbitrary but operate within physical law. Soul-actions are physicochemically apposite because they make sense in a physical world. Understanding the actions of souls requires explanations of motivations and meanings as well as of mechanisms. Hermeneutics and biology coexist in reciprocal tension. Biology explains the evolutionary and developmental origins of interpretative souls. Their understanding is matter for interpretation.

What Is It Like to Be a Slug?

> In this new world they no longer possessed their former guides, their regulating, unconscious and infallible drives: they were reduced to thinking, inferring, reckoning, co-ordinating cause and effect, these unfortunate creatures; they were reduced to their "consciousness," their weakest and most fallible organ!
> —Friedrich Nietzsche (2007)

The relation of parts to wholes was one aspect of Dilthey's *Verstehen*. Another was subjectivity: the direct access we have to inner experience and how this relates to our understanding of other subjectivities and the objective world. All meanings exist for an interpreter, but introspection protests that the subjectivity of riboswitches is strictly metaphorical. We strongly doubt there is anything it is like to be a riboswitch. We have no empathy *for* riboswitches. But is there something it is like to be a chimpanzee or a slug? We have more empathy for chimpanzees than for slugs because we find it easier to put ourselves in a chimpanzee's shoes. It is harder to imagine a slug wearing shoes, but my better acquaintance with slugs suggests that they too have some form of subjective awareness. When two slugs entwine in a nuptial embrace, each both sexes in one, I conjecture a frisson of joy.

The aspect of human souls that we know from empathic experience is spoken of as conscious awareness. *Consciousness*, *awareness*, *attention*, *concentration*, and *engagement* are interrelated terms for what we perceive to be limited resources of functioning souls. The *Oxford English Dictionary* is of limited help. *Conscious* is derived from a Latin root "to know" and, like the cognate word conscience, has connotations of guilt. *Aware* comes from an Old English word for "cautious," the latter from a Latin root for "to beware." *Attend* comes from a Latin root for "to stretch," cognate with tension. Concentration comes from a Latin root for "to bring to a common center." *Engage* comes from a French root for "to pledge." These etymologies suggest associations with knowing, danger, direction, responsibility, centrality, and commitment.

Most problems are solved without consciousness. Our bodily actions involve myriad decisions at many spatial and temporal scales. We lack conscious access to most bodily choices, although we may be conscious of their downstream effects. For present purposes, I will distinguish autonomic and automatic actions while acknowledging no clear-cut distinction between them. Autonomic actions are fully unconscious, whereas automatic actions can, when challenged, be brought under conscious control. Examples of autonomic actions include the beating of my heart, the inspiration of breath, the influx and efflux of ions across membranes, and the construction of consciousness. Examples of automatic actions include repeated behaviors such as the scratching of an itch and the lean of a cyclist's body as she rounds a bend. It is a common subjective perception that conscious engagement with a task plays a role in some forms of learning, but that once complex behaviors have been learned, many are better performed automatically than consciously.

The mechanisms of consciousness are tools, and the contents of consciousness are texts, used in higher-level interventions into lower-level mechanisms; but the implementation of all interventions depends on a substratum of autonomic and automatic difference making. When I walk to work, all of my cellular choices are autonomic and many of my higher-level choices are automatic. I am unaware of complex decisions where to place my feet and when to turn a corner. Freed from these mundane tasks, my consciousness is gainfully employed in planning my day or wanders freely among idle thoughts, but my attention becomes highly engaged in crossing major roads. On days when I intend to deviate from my usual route to go to the dry cleaners, I need to keep this intention near the forefront of my attention or I will arrive at the office with my dirty laundry still clutched in my hand. My phenomenological walk to work is very different on icy winter mornings when decisions where to place my feet are at the forefront of consciousness. A brief lapse of attention may cause a damaging fall. Multiple resources are brought to bear on the problem of foot placement. It is a high priority. When a slug breaks cover to venture forth across a paved path, it is alert for danger. It does not know where it is heading. Or perhaps it knows because it has been there before.

Phenomena are interpretations of peripheral inputs. They are metaphors of things in the world used to inform choices of action. In the background of my more or less conscious awareness, a perceptual model of the world is continually updated with new data. New inputs are compared to interpretations of past inputs. Model and data are brought into register (in the printer's sense of bringing into alignment). Percepts function as detectors of difference between model and data. Alarming differences are flagged as deserving immediate attention.

The contents of consciousness are interpretations ready for use. They include items for which I do not need to search and pointers for where to search for more information. My perceptual field categorizes salient objects, situates them in space, and connects them to a storehouse of knowledge and my intentions toward them. When I needed a pen to write the first draft of this paragraph, I did not start from ignorance as to where a pen might be. I was vaguely aware without looking that a pen was by my right hand. I then turned my attention to pick up the pen. I did not need to start from scratch by searching for a dark cylindrical object, that might or might not be there, and then engaging in further interpretation to decide whether the cylindrical percept was a pen or a laser pointer. All that mental work had already been done. But before I decided to pick up the pen, its location was not at the forefront of my consciousness but in a nebulous background of content that was "at hand" if needed. I knew where to look.

Consciousness includes "holding in mind," a short-term memory that enables coordination and coherence of actions. I recently had the intriguing nonexperience of suffering septic shock. I was seemingly alert, answered questions when asked, but gave the same answers over and over again. This is what I have been told. I seemed to observers to be conscious and probably would have said I was conscious if asked—what a silly question!—but unlike my wife who attended with care I had no recollection of what I just said. I was temporally incoherent. I remember nothing between being put in the ambulance and "coming to myself" when recollection returned as autonomic processes of cellular souls reestablished higher-level coherence. "I" came back.

A perennial "hard problem" of philosophy is how subjective awareness exists in a material world. I have no particular insight into this problem but suspect that progress will be made by gaining better understanding the final causes of consciousness, the tasks it has evolved to solve. Consciousness is a tool used as a private text. It is an interpretation of information that is used by the soul to inform subsequent interpretations. What features distinguish the tasks that require this textual prosthesis? Why cannot these tasks be performed as effectively autonomically or automatically? What are the material media on which the text is inscribed? I suspect that answers to these questions will require entering into the hermeneutic circle in which complex higher-level "considerations" intervene on lower-level mechanisms that are the objective substrate of the higher-level subjectivity. Consciousness requires engagement with metaphor.

Objective Phenomena

> The self-awareness of the individual is only a flickering in the closed circuits of historical life. *That is why the prejudices of the individual, far more than his judgments, constitute the historical reality of his being.*
> —Hans-Georg Gadamer (1992)

A perennial not-so-hard problem of philosophy is how objectivity can exist in a phenomenal world. There is a sense in which all knowledge is subjective, with nothing known outside of perception. An organism does not have direct access to things in themselves, only to interpretations of things, but the interpretative faculties of organisms have evolved to enable effective action in the world and, for this reason, can be relied on to provide useful guidance. Our verdicts are just. An organism brings past prejudices—"the historical reality of its being" in Gadamer's phrase—to all present judgments. Objective "facts" are things upon which we all can agree. Rough consensus exists among members of a species about the nature of worldly things because their shared evolutionary history has given them similar sensory and interpretative mechanisms. They have the same prejudices.

Kant wrote: "What the things may be in themselves I do not know, and also do not need to know, since a thing can never come before me except in appearance" (1781, 375). So much for pure reason! One practical reason why one might need objective information about a thing is that it will eat you if you do not take evasive action. A gazelle is eaten by the cheetah *an sich*, and becomes flesh of its flesh, rather than appears to be eaten by an appearance. Subjectivity is objectively grounded in the need to know.

When we interact with inanimate things, we act most effectively when we see the world "as it is." We can trust our perceptions because inanimate things do not have purposes. Our interactions with animate things have a different flavor. We cannot always accept what we see at face value. They may have purposes that conflict with our purposes. They may have incentives to hide their intentions and manipulate our perceptions, just as we have incentives to hide our intentions and manipulate their perceptions. The evolution of our perception drives the evolution of their deception. We perceive most effectively when we "see through" their stratagems, but we deceive most effectively when we hide our motivations, perhaps even from ourselves, if self-deception gives less away to their heightened abilities to detect our deceit.

Subjectivity is an objective attribute of organisms. For this reason, an objective understanding of organisms, including ourselves, requires engagement with their subjectivity. We have evolved objective capabilities of second-person and third-person sympathy that allow us to model the subjective perspectives of others and anticipate how they will react to our actions, and these sympathetic abilities help us gain a more objective understanding of ourselves. But we distrust the objectivity of others, because human agents play games, including with evidence. We should likewise distrust our own objectivity because we are unjust like them. Our perceptions of our own motivations and of what is "fair" may be biased to favor our own ends. We have self-interested incentives to present our preferences as disinterested. For human agents who articulate their reasons, a readiness to see value in the other side of an argument can place one in a weaker position in bargaining with an adversary who sees only their side. I recognize within myself an ability to quarantine my objective judgments of multiple perspectives from my passionate conviction that the "objective" evidence supports only my side.

Finally, our perceptions may be distorted because we are used as means to the ends of other agents, both external and internal. How we perceive the "objective" world can be shaped by external actors, using social media, and by internal actors, using emotional media, to manipulate what we perceive. Our self-perceived interests may not serve our self-professed ends. Our hopes and our joys, our fears and our hatreds, have evolved to serve our genes' replicative ends. We have evolved to be ill-satisfied, endlessly desiring something more. We do not rest on our laurels. The pursuit of happiness is an addiction to striving. The promise of contentment when we get what we want is often false advertising.

A trope embraced by many scientists is that science is objective. Humanists have their own well-developed criteria for weighing evidence and judging the value of competing interpretations, and they detect, in this trope, an invidious comparison that valorizes "hard" over "soft" scholarship. Scientists and humanists have their own subjectivities and objectivities, and I do not wish to enter into these arguments. Rather, I wish to suggest that biologists need to incorporate subjectivity into their objective understanding of living things and may have something to learn from the humanities on the subject.

Telling Tales

If one cannot recall everything, neither can one recount everything. The idea of an exhaustive narrative is a perfomatively impossible idea. The narrative necessarily contains a selective dimension.
—Paul Ricoeur (2004)

Historical narratives are created rather than discovered. They are fruits of Minerva and the Muses, not Daedalus and Vulcan. They are interpretations of the past that require further interpretation before application to present concerns. Useful histories are not exhaustive catalogs of everything that happened but attempts to identify events and patterns that are of particular significance. They are a search for *reasons why* that requires judgment and discrimination. A century after Dilthey, Ernst Mayr, in his final book published in his hundredth year, wrote:

With the experiment unavailable for research in historical biology, a remarkable new heuristic method has been introduced, that of *historical narratives*. Just as in much of theory formation, the scientist starts with a conjecture and thoroughly tests it for its validity, so in evolutionary biology the scientist constructs a historical narrative, which is then tested for its explanatory value. . . . Evolutionary biology, as a science, in many respects is more similar to the Geisteswissenschaften, than to the exact sciences. When drawing the borderline between the exact sciences and the Geisteswissenschaften, this line would go right through the middle of biology and attach functional biology to the exact sciences while classifying evolutionary biology with the Geisteswissenschaften. (2004, 23)

The scientific critics of adaptationism have it right. Evolutionary narratives do not conform to the norms of hard science. They have more in common with the methods of the *Geisteswissenschaften*. Stephen Jay Gould, wearing his structuralist hardhat, derided functionalist narratives:

Rudyard Kipling asked how the leopard got its spots, the rhino its wrinkled skin. He called his answers "just-so stories." When evolutionists study individual adaptations, when they try to explain form and behaviour by reconstructing history and assessing current utility, they also tell just-so stories. . . . Virtuosity in invention replaces testability as the criterion for acceptance. . . . When we examine the history of favored stories for any particular adaptation, we do not trace a tale of increasing truth as one story replaces the last, but rather a chronicle of shifting fads and fashions. (1980, 259)

In a broadside against Panglossian adaptationism, Gould and
Lewontin (1979) valorized the strictures of structure as constraints on evolutionary freedom but, when Gould (1990) came
to recount the story of the Cambrian explosion, his narrative
now favored unpredictability (contingency) over inexorabil-
ity (destiny): "The resolution of history must be rooted in the
reconstruction of past events themselves—in their own terms—
based on narrative evidence of their own unique phenomena"
(278). He objected not to storytelling *per se* but to particular stories being told.

Adaptationists aspire to tell the story of the synchronic *what
for?* not just the diachronic *how come?* The time of adaptationist
narratives is not a teleological time with a preordained end nor
a directionless time of pure contingency. It is a recursive time of
repetition with variation, in temporal cycles of birth unto birth
and the copying and re-copying of genetic texts. The textual
record of what worked in past performance permits the persistence of tradition into the future. Some stories are better than
others.

Historical narratives arouse passions because what is emphasized and what is erased can be perceived as promoting partisan
penchants of the present. Can there be an objective history? Historians have grappled with this question since at least the nineteenth century. There are historical facts, on which historians
can agree, and interpretations of these facts which are subjects of
dispute. Historians have developed standards for handling evidence and moving forward in understanding. By general agreement, some histories make a stronger case than others for their
interpretation of the past, but *what has been* is always subject to
reinterpretation. Evolutionary historians can do no better.

Beyond Natur und Geist

"How *could* anything originate out of its opposite? for example, truth out of error? or the will to truth out of the will to deception? or selfless deeds out of selfishness? . . . The things of the highest value must have another, *peculiar* origin—they cannot be derived from this transitory, seductive, deceptive, paltry world, from this turmoil of delusion and lust."

—Friedrich Nietzsche (1886, in ironic voice)

The moat between science and the humanities can be deep. From the side of the natural sciences, humanists' obsession with form over substance, their focus on values rather than facts, their neglect of physical causation, can be viewed as vacuous impediments to effective interventions to improve human well-being. From the side of the humanities, scientists' obsession with parts over wholes, their focus on facts rather than values, their neglect of cultural causation, can be viewed as soulless impediments to the amelioration of human needs. Efficient and material causes are the wards of Daedalus and Vulcan. Formal and final causes are assigned to the care of Minerva and the Muses. We have come a long way from the scholastic consensus that understanding something required attention to all four Aristotelian causes. But one can admire what one sees on the other side of a ditch and welcome an exchange. The heavy hand of Vulcan may crave the lighter touch of Clio or Terpsichore.

The central claim of this book is that formal and final causes arose from efficient and material causes by historical processes. I believe that the *dogmatic* exclusion of teleological considerations from the working philosophy of most biologists has become an impediment to scientific progress. I believe that organisms, sometimes even material genes, interpret their world in meaningful ways. I believe that meanings and values have been present from the very origins of life, although these were not human meanings nor human values. I believe that the boundary between meaningful lives and mere existence should be placed between the living and nonliving worlds. I believe that the scientific rejection of a naturalized teleology has contributed to the estrangement of scientific and humanistic modes of understanding. I believe that the humanities and social sciences have insights to contribute to biology about questions of meaning, value, and interpretation. I believe that a Darwinian account of the origins of meaning has something to say to the humanities and social sciences about how our unique abilities have emerged from features we share with other living things. Here I stand. What shall I do next?

I would like to believe that a fuller understanding of the implications of natural selection for questions of meaning and purpose could lessen the culturally constructed barriers between the *Naturwissenschaften* and *Geisteswissenschaften*, but the prospects of natural selection reconciling Minerva and Vulcan do not, at first sight, appear promising. Darwinian accounts of human nature remain a lightning rod for all those who wish to maintain a separation between domains of physical inquiry and meaning. For religious fundamentalists, Darwinism has been uniquely singled out as the enemy within soulless science because of its claims that means can be fitted to ends without a directing intelligence. Creationists see Darwinism as teleology "shorn of all its goodness." Other scientific disciplines are equally incompatible with scriptural literalism, but none arouses comparable religious opprobrium. From within the *Geisteswissenchaften*, Darwinian hypotheses about human nature provoke censure as illegitimate and puerile encroachments of simplistic science into domains where it does not belong. Even within evolutionary biology, adaptationist explanations are disdained as just-so stories, and theoreticians who emphasize the role of neutral processes consider themselves more rigorous than those whose principal interest is adaptive fit to the conditions of existence (I know this from personal interactions with faculty in my own department).

Adaptationism is singled out for disapprobation *because* it reaches across a boundary. On the physicalist side, adaptive "storytelling" is seen as polluting the pristine province of efficient and material causes with the messiness of meaning. On the humanist side, Darwinian explanations of human nature are rejected as hostile takeover bids by advocates of mindless mechanism. Both physicalists and humanists are happy with the border where it currently stands (we are not like them!). Territorial borders are historically contingent obstacles to freedom of movement across a continuous terrain. In terms of academic *Realpolitik*, Darwinism resides in the borderlands of *Natur* and *Geist*, a small principality wedged between hegemonic powers.

Many scholars have a visceral dislike of Darwinism. Their reactions relate to the four aspects of random variation, survival of the fittest, agency, and determinism. The role of chance is perceived as positing a world without meaning; natural selection is perceived as bleak and harsh; the ascription of purpose to other organisms, even genes, is perceived as blind to the uniqueness of human agency; and the explanation of human nature by natural processes is perceived as denying our freedom to restructure our world as we choose. It does not help to point out that natural selection is simultaneously criticized for invoking a world without meaning and for finding meaning where it does not belong.

The easiest concern to address is human uniqueness. Every species is unique, but humans possess culture and language to a degree that far surpasses any other species. We are a symbiosis of genes and memes that can adapt to a changing world much faster than the measured pace of genetic change. I have not addressed the extraordinarily complex and subtle mechanisms of the human brain, nor the dialectics of minds in societies, by which we make sense of our worlds. That is work for others. My limited goal has been to find continuity, rather than discontinuity, between the simplest and most complex forms of interpretation. I also hope to have lessened concerns about determinism. We have uncounted, although not infinite, degrees of freedom in observation and action. We are self-motivated agents that are buffered from external causes. We can stand our ground as unmoved movers of the here and now.

Many critics of Darwinism cannot see how directionless chance could generate something as complex as a living thing. But chance does not act alone. Natural selection preserves the progeny of fortunate accidents and the progeny of those progeny with additional fortunate accidents, while it eliminates progeny with unfortunate accidents and those without recent serendipities. The difference between a clerical error and a fortunate slip of the pen—between a false note and nailing it—is that the serendipitous error is retrospectively endowed with meaning once it is copied and recopied. And so it proceeds like a ratchet. This is the creative power of natural selection. The meanings of life bear the trace of what they are not.

Another common concern is that the role of chance seems to remove meaning from life. We are here for no reason. But chance does not act alone. Natural selection produces beings with reasons for being. It is the amoral architect of human beings capable of moral and immoral choices. "The ends justify the means" could be a definition of natural selection once "justify" is stripped of any connotations of moral approval. Natural selection has a bright and a dark face. They are two sides of a coin. The bright face reflects all the beauties and exquisite adaptations we find in the living world. The dark face hides the selective culling of the less fit in lives that are often nasty, brutish, and short. The bright face is born out of the dark face and is sustained by the dark face. The combination of beauty and cruelty is unsettling. The interplay of light and shadow is the *pathos* of life. We make of our lives what we will.

Cadenza

The ascription of utilitarian purposes to natural things leaves many cold. It is not how they react to "works of nature." They prefer an aesthetic response. I will end with the sestet of one of Shakespeare's great sonnets:

Yet in these thoughts my self almost despising,

Haply I think on thee, and then my state,

Like to the lark at break of day arising

From sullen earth, sings hymns at heaven's gate;

For thy sweet love remembered such wealth brings

That then I scorn to change my state with kings.

The lark ascending has inspired the poetry of William Shakespeare, Percy Bysshe Shelley, and George Meredith, and the music of Ralph Vaughan Williams. The display of the male skylark has been used to evoke selfless joy, but a modern Darwinist might describe the display as an honest, because strenuous, signal of quality made credible by the risk of being taken by a hawk. The lark is probably not bursting with joy, but exhausted and afraid. Shakespeare's and Shelley's transcendent metaphors could be seen as reflecting an erroneous view of the natural world. Does the striving of the skylark have no higher end than showing off to attract mates? Is there squalor in this view of life? Must Darwinists live in a disenchanted world? When I observed a lark at break of day arising from an Oxford meadow in cascades of song, did the richness of the scientific vision impoverish the poetic image?

I saw again the skylark's flight,

Rising up on Solsbury Hill,

And once again my spirit rose,

My heart pounding with the climb.

Appendix (a Vestigial Organ): Words about Words

> Rudimentary organs may be compared with the letters in a word, still retained in the spelling, but become useless in the pronunciation, but which serve as a clue in seeking for its derivation.
>
> —Charles Darwin (1859)

When I query an informant about what a word means, I frame the question as a spoken or written linguistic text and am answered with a linguistic text. We communicate meanings as words. For many philosophers, questions of meaning are primarily questions about language, but this book has generalized the concept of meaning to interpretations of all kinds. Linguistics and philosophy of language are vast territories of erudition into which an ill-informed novice should venture with trepidation. Nevertheless, I have been persuaded that I need to at least sketch how I would relate my account of meaning to language.

An important reason to read a text is to understand the author's intentions. And an important reason why an author might compose a text is to have her intentions understood. Languages are elaborate conventions, shared by communities of speakers, used for the composition and interpretation of linguistic texts. Conventional meanings evolve because authors and readers often both benefit from mutual understanding of authorial intentions, but the difference between an author's intentions and how the author intends her text to be interpreted means that language can be used to misinform as well as inform.

I have defined *meaning* as the physical output of a process of interpretation and *text* as an interpretation intended to be interpreted. By these definitions, two kinds of text are central to language that I will call *public* and *private texts*. Public texts are the strings of spoken or written words that are outputs of language users and that are inputs perceived and interpreted by other language users. Each language user possesses a private text used in the composition and comprehension of public texts. The private text has an intricate material form. It is a text because (1) it is a physical interpretation of the language user's life-experience in the context of innate *a priori* knowledge, and (2) it informs the composition and comprehension of public texts. The private text is informed by the evolutionary and developmental history of the language user, especially her lived experience of public texts. Mastery of a language is the context that allows decryption of texts written or spoken in words of that language. A child learns a language in much the same way as British naval intelligence (Room 40) broke German code 7500: by observation of many exemplars together with inspired guesses, tested for intelligible meaning in encounters with multiple texts, all in the context of innate and learned knowledge of how humans think. This is the hermeneutic circle.

My public texts are roughly synonymous with Saussure's (1916) *parole* and my private texts with his *langue*. Saussure emphasized the communal nature of *langue* but I emphasize the personal nature of *private texts*. Private texts are unique to each user, but the needs of mutual understanding result in mutual information, and convergence on shared conventions, among the private texts of a linguistic community. Private texts develop over the course of a life from the cumulative perception and interpretation of public texts in linguistic and nonlinguistic context. The forms of public texts are arbitrary conventions unique to each language, but the initial bootstrapping—which allows a private text to be informed by public texts—requires key inputs from genetic texts.

Public texts are designed to evoke rich associations in the private texts of intended auditors, just as artfully placed splodges of an impressionist painting evoke interpretive mechanisms in viewers that "fill in the details." The richness of information resides in the auditors' private texts, not in the public texts that function as conventional pointers to contents of private texts. This indicative function of public texts depends on communicants sharing similar-enough private texts because of common humanity, similar life experiences, and shared membership in a linguistic community (they need to speak the same language). A speaker anticipates that a listener will interpret the public text in much the same way as the speaker would interpret the public text.

Words are defined using other words. For the writing of this paragraph, I opened a Magyar dictionary at random and found the word *csendülni* followed by other words that I assumed to be a definition of *csendülni* in Magyar. The public text *csendülni* pointed to nothing in my private text. It evoked no associations in my mind. If you are not a Magyar speaker, there is probably little you can say about *csendülni* except that you believe it to be a word of Magyar on my dubious authority. Most Magyar speakers could probably use *csendülni* in conversation, but it would have different private associations for each speaker. My friend Apari Péter offered "voice of the bell" as his spontaneous translation of *csendülni*. Some Magyar speakers might have no more than a vague idea what the word "really" means and might consult a Magyar dictionary for the "correct" definition. When I consult an English dictionary, I am not infrequently surprised to find that the public definition differs markedly from the sense of my private text.

This public text is an attempt to persuade; to rearrange the associations of *meaning* and *information* in your private text; to change how you interpret and use these words not reveal what they "properly" mean. It is an attempt to *explain* so that you will *understand* my intention when you consult your private text. My intent is to construct abstract attractors in your private text that will give form to your future thought.

Supplement to the Appendix

> Think of words as instruments characterized by their use, and then think of the use of a hammer, the use of a chisel, the use of a square, the use of a glue pot, and of the glue.
> —Ludwig Wittgenstein (1958)

Dilthey (1883) distinguished the *Naturwissenschaften* (natural sciences) as sciences of *Erklären* (explanation) from the *Geisteswissenschaften* (spiritual sciences) as sciences of *Verstehen* (understanding). For Ricoeur (1971), the dialectic of explanation and understanding was the hermeneutic circle ("I *understand* what you intended to do, if you are able to *explain* to me why you did such-and-such an action"). Understanding is often used with connotations of recognition of an intention. An author considers a reader to have understood his text when the reader's interpretation is close to that intended by the author and the author believes he has been misunderstood when the reader's interpretation differs markedly from the author's intended interpretation. A reader understands where the author "is coming from" when the reader recognizes the author's intention. Authors and readers often disagree about whether the text has been understood.

In all human discourse there is ineliminable freedom of interpretation. Consider translation of a Latin text into English by a human interpreter. The inputs to translation are of two kinds: *text* and *context*. The output is a text in English. Different translators produce different translations of the same text because of differences of personal context. The Latin text does not *dictate* the English translation. There is, in principle, an extraordinarily complex synchronic account of the molecular mechanism by which the Latin text is translated as an English text given the molecular state of the translator, but an explanation of this molecular state depends on the even more complex evolutionary, developmental, and cultural history of the translator (what one might call the diachronic sources of the translator's personal identity or personality).

Definition is use. A word can denote subtly different things for different speakers within a linguistic community. Some speakers may have completely "wrong" definitions because they have *misunderstood* what they have heard. Some variants leave few descendants, perhaps they are used and corrected in the classroom, or the word itself falls into disuse because denoted things disappear from the environment. Other variants spread though the community and the "wrong" definition becomes standard usage. The survival value of a definition is determined by its role and use within the linguistic community. Some words are "living fossils" whose definitions change slowly or not at all, whereas other words possess rapidly changing definitions. The meanings of *gene* and *gender* have changed dramatically during this century, with several usages currently in competition. Some older usages may disappear and be replaced by new words as their proponents become tired of being misunderstood.

Each speaker's association between a word and something in the world is a personal convention, but the convention must have survival value within the community of speakers if it is to persist. The attributes that favor survival of a personal convention are various, but two can be singled out. First, the convention is more likely to persist if similar conventions are possessed by other speakers and listeners. Second, a convention is more likely to persist if the word corresponds to some significant property of the world. By some such process, gestures and sounds (or their subvocalizations) came to be *about* things in the real world. Origin, function, causes, and consequences are inextricably entwined in recursive systems.

The definition of a word is an evolving entity with a history but without essential attributes. Within the philosophy of biology there has been a shift from defining species as *classes* to defining them as *individuals*. An organism belongs to a particular species by virtue of its ancestry rather than by the possession of defining features. From a diachronic perspective, words have the same kind of individuality. But, from a synchronic perspective, word tokens are members of classes with consensus definitions. Words are historical kinds.

Languages are technologies of the self that have evolved as elaborate systems of reciprocal altruism to allow humans to coordinate their activities to avoid dangers and exploit opportunities. But languages are also used by language users to exploit other language users. What distinguishes morally legitimate means of persuasion from morally questionable manipulation and immoral coercion? A full exploration of this question is beyond my current intent and I will limit myself to mentioning two important factors. One criterion of morally questionable means is the imposition or threat of large costs if another individual does not make a desired choice. Another criterion is deceit: hiding from the individual you are attempting to influence the reasons for your intervention.

Supplement to the Supplement

> Es ist heute unmöglich, bestimmt zu sagen, warum eigentlich gestraft wird: alle Begriffe, in denen sich ein ganzer *Prozeß* semiotisch zusammenfaßt, entziehen sich der Definition; definierbar ist nur Das, was keine Geschichte hat.
> —Friedrich Nietzsche (1887)

> Today it is impossible to say precisely why people are actually punished: all concepts in which an entire *process* is semiotically concentrated defy definition; only something which has no history can be defined.
> —Friedrich Nietzsche (2007)

German *Prozeß* has a secondary meaning of criminal trial that eludes translation by the English *process*.

There are two general ways of persuading readers to describe and interpret the world in your terms. One is to persuade them to adopt your definitions of existing terms. Another is to persuade them to use new terms of your invention. In this supplement of the supplement, I consider successful redefinitions of *gender* and, in the supplement hereto, the forlorn failure of *madumnal*.

In the first chapter of *David Copperfield*, Charles Dickens (1849) wrote: "In consideration of the day and hour of my birth, it was declared by the nurse . . . first, that I was destined to be unlucky in life; and secondly, that I was privileged to see ghosts and spirits; both these gifts inevitably attaching, as they believed, to all unlucky infants of either gender, born towards the small hours on a Friday night." *Gender* has long been used as a synonym for *sex*, although this sense was described as "now only jocular" in the first edition of the *Oxford English Dictionary* (1899). At that time, *sex* was the usual term for the distinction between males and females, but, within a century, *sex* was replaced by *gender* as the preferred term of many English speakers.

I have studied recent shifts in the use of *sex* and *gender* as a test case for thinking about memetic evolution (Haig 2004b). In brief, social psychologists and psychoanalysts introduced a distinction between "socially constructed" gender and "biologically determined" sex in the 1960s. *Gender* became a term of art in these fields, both as a way of marking a theoretical distinction and of signaling to informed listeners that the speaker believed the social to be more important than the biological. Animals had sex; only humans had gender. From this base, *gender* entered general discourse via its adoption by feminists of the 1980s to signal their belief in the predominant influence of social factors on sex differences, but a sex–gender distinction is now rarely maintained. *Gender* commonly now refers to all differences between males and females whether these are social or biological. Even hamsters have genders (Robins et al. 1995).

Haig (2004b) suggested two factors played a role in the reconvergence of meanings of *gender* and *sex*. First, biological and social factors often interact to determine differences between men and women, and, in such situations, there was no neutral term. *Gender* became the safer default in cases of overlap, thus undermining the distinction between biological sex and cultural gender. Second, many listeners heard *gender* being used to refer to male–female differences but were uninformed of the theoretical distinction and thus adopted *gender* as the fashionable general term. I now suspect another factor contributed to the rise of *gender* and concomitant decline of *sex*. As a student explained her preference for gender to me: "Gender is a category but sex is an action." Some speakers avoid using *sex* because *sex* has copulatory connotations. *Sex* became a widely used euphemism for fucking only during the twentieth century. Did you wince when you read the "F-word"? I hesitated as I wrote it and still hesitate when I read it. Euphemisms are polite ways of saying things we feel uncomfortable saying. But, as a euphemism becomes widely adopted, our discomfort with the subject inevitably rubs off on the euphemism, which becomes less polite (consider the linguistic fate of "toilet," now strongly associated with the "S-word"). Our hesitance in talking openly about "sexual activity" has tainted *sex* with "sexual connotations" and created a preference for a word in which these associations are less direct. Oddly, an obsolete definition of *to gender* is to copulate.

The public forms of sex and gender, as spoken and written texts, as vibrations in air and marks on paper, have remained unchanged through substantial changes of usage in context. Change has occurred in the private associations of the public forms. Meanings of words have histories not essential attributes. At any one time, different speakers would have given different interpretations of what the public forms meant and used the public forms in different contexts. If enough speakers use a word "wrongly" then their private definition becomes the accepted norm of their linguistic community. The heterodox becomes orthodox. Meaning is use.

Supplement to the Supplement to the Supplement

> Across this sequence of supplements a necessity is announced: that of an infinite chain, ineluctably multiplying the supplementary mediations that produce the sense of the very thing they defer.
>
> (citation deferred)

An example of failed neologism is my attempt to persuade workers in the field of genomic imprinting to recognize a distinction between *maternal* alleles (*in* the mother) and *madumnal* alleles (*from* the mother in the offspring), with a parallel distinction between *paternal* and *padumnal* alleles. A mother possesses two alleles at each locus, one of which is transmitted to her offspring via an egg. The adjective "maternal" is commonly used to describe *both alleles* of the mother and a *single allele* of the offspring. These two uses of the adjective describe different situations that are subject to distinct selective forces. As an example of confusions that can arise, consider an imprinted gene that is expressed only from the allele that a daughter inherits from her mother that affects the maternal care that the daughter provides to her offspring. To which material genes would *maternal gene* refer in this case? Greater clarity could be achieved if one had some succinct way of distinguishing a gene inherited "from a mother" from a gene "in a mother." I needed a linguistic difference to mark this semantic distinction.

My first solution was to use *maternal* for "in the mother" and *maternally derived* for "from the mother" (Haig and Westoby 1989), but *maternal* continued to be used in the scientific literature for both senses of *in* and *from* the mother. Readers considered the final three syllables of "maternally derived" to be superfluous. I next proposed a distinction between *maternal* (in mothers) and *madernal* (from mothers) (Haig 1992a). This had a certain minimalist elegance, but no one adopted *madernal* and the distinction between *maternal* and *madernal* was easy to miss in spoken English (there is no discernible difference in accents with soft t's). I next proposed *maternal* (in a mother) and *madumnal* (from a mother) (Haig 1996b) the latter term modeled on *autumnal*. This neologism had the advantage that readers and listeners were forced to note the difference in pronunciation of *maternal* and *madumnal*, but the disadvantage that *madumnal* was almost universally unloved. Was this because *madumnal* was intrinsically cacophonous or because the unfamiliar was perceived as uncomely and might be warmly embraced with better acquaintance? If my ugly duckling were to transform into a graceful swan, then you would be seeing the world more nearly as I see it. The most important reason for the failure of my terminological innovation was that few readers saw the *need* for such a distinction. I am stubborn and still think the distinction is important but, for this volume, I have returned to the use of *maternal* and *paternal* because, on balance, I decided that removing the distraction provided by unfamiliar terms was worth the resulting loss of precision.

All meaning is metaphor. One thing stands in the place of another. A spoken word is transcribed as a written word that is translated as a written word in another language. The signifier is a metaphor of a signified that has been a metaphor of a signifier. The *significance* of texts is that texts perpetuate past choices for guidance of future choice. The interpretation of interpretations is endless. No one has the final word.

Acknowledgments

It is a common human failing (to which I am susceptible) of emphasizing our own contribution to what goes well and finding extenuating circumstances for our participation in what ends poorly. I have found that the more I read, the less original I become, not in the sense intended by the sign on the wall of Francis Crick's and Sydney Brenner's laboratory "Reading rots the mind" but in the sense, that if I read broadly enough, I find that all of "my" ideas have occurred to somebody else. Who is to say whether these are independent inventions or whether there is some complex historical connection. Therefore, thanks to the many unacknowledged influences without whom this book would not have been written in quite the way it was. Ideas are communally generated, and this book was no exception. It is a reformulation of things I have read and things I have heard from many people, and to them, even though they are not otherwise acknowledged, thanks. Special thanks are due to Daniel Dennett, a friend of many years, without whom it definitely would not have been written, and to the participants in the Colloquium on the Fundamental Interconnectedness of All Things, who constantly challenged me to question my assumptions.

Peter Apari, Maarten Boudry, Alex Byrne, Richard Bondi, John Brockman, Rosa Cao, Kathleen Coleman, Bernard Crespi, Helena Cronin, Nick Davies, Daniel Dennett, Patrizia d'Ettorre, Thomas Dickins, Holly Elmore, Steve Frank, Andy Gardner, Shantanu Gaur, Peter Godfrey-Smith, Alan Grafen, David Hughes, Laurence Hurst, Justin Jungé, Elias Khalil, John Krebs, Jeffrey Lipshaw, Avantika Mainieri, Dakota McCoy, Jeff McKinnon, Latha Menon, Lucas Mix, Pavitra Muralidhar, Eric Nelson, Saul Newman, Martin Nowak, Samir Okasha, Manus Patten, Naomi Pierce, Jura Pintar, Susannah Porter, Robert Prior, Barbara Rice, Judith Ryan, Richard Schacht, Eric Schliesser, Karl Sigmund, James Simpson, Adam Smith, Adam Smith, Stephen Stearns, Kim Sterelny, Richard Thomas, Wenfei Tong, Robert Trivers, Francisco Úbeda de Torres, Carl Veller, Helen Vendler, Brianna Weir, Florence Westoby, Mark Westoby, Jonathan Wight, Jon Wilkins, Jason Wolf, and Adrian Young are all authors of this book.

And thanks to the rejectors of manuscripts who forced me to seek a better venue. You are unwitting authors of this book. Thanks to the choices of my ancestors that were, a few generations back, the choices of your ancestors, and to my daughter, Jesse Gudrun, who makes life worth living, and my sons, Cesar Aeneas and Liam Ulysses, wanderers in the wide world. Especial thanks to Eneida Pardo and a long-dead butterfly in Marilia who knew not what it did.

References

Abramowitz, J., D. Grenet, M. Birnbaumer, H. N. Torres, and L. Birnbaumer. 2004. XLαs, the extra-long form of the α-subunit of the Gs G protein, is significantly longer than suspected, and so is its companion Alex. *Proceedings of the National Academy of Sciences USA* 101:8366–8371.

Ackrill, J. L. 1973. Aristotle's definitions of "psuche." *Proceedings of the Aristotelian Society* 73:119–133.

Adami, C. 2002. What is complexity? *BioEssays* 24:1085–1094.

Adami, C., C. Ofria, and T. C. Collier. 2000. Evolution of biological complexity. *Proceedings of the National Academy of Sciences USA* 97:4463–4468.

Adams, D., K. Horsler, and C. Oliver. 2011. Age-related change in social behavior in children with Angelman syndrome. *American Journal of Medical Genetics Part A* 155:1290–1297.

Ainslie, G. 2001. *Breakdown of Will.* Cambridge: Cambridge University Press.

Alexander, R. D. 1987. *The Biology of Moral Systems.* New York: Aldine de Gruyter.

Alvares, R. L., and S. F. Downing. 1998. A survey of expressive communication skills in children with Angelman syndrome. *American Journal of Speech-Language Pathology* 7:14–24.

Amundson, R., and G. V. Lauder. 1994. Function without purpose: The uses of causal role function in evolutionary biology. *Biology & Philosophy* 9:443–469.

Aquinas, T. 1965. *Selected Writings of St. Thomas Aquinas*. Indianapolis: Bobbs-Merrill.

Aquinas, T. 1975. *Summa contra gentiles*. Book 1: *God*. Notre Dame: University of Notre Dame Press.

Arima, T., T. Kamikihara, T. Hayashida, K. Kato, T. Inoue, Y, Shirayoshi, et al. 2005. *ZAC, LIT1 (KCN1Q1OT1)* and *p57^{KIP2} (CDKN1C)* are in an imprinted gene network that may play a role in Beckwith-Wiedemann syndrome. *Nucleic Acids Research* 33:2650–2660.

Aristotle. 1984. *The Complete Works of Aristotle*. Princeton: Princeton University Press.

Arrow, K. J. 1963. *Social Choice and Individual Values*. 2nd ed. New Haven: Yale University Press.

Atkins, P. W. 1994. *The Second Law*. New York: W. H. Freeman.

Aughton, D. J., and S. B. Cassidy. 1990. Physical features of Prader-Willi syndrome in neonates. *American Journal of Diseases of Children* 144:1251–1254.

Augustine of Hippo. 2012. *City of God*. Part 1: *Refutation*. Trans. M. Dods. London: Folio Society.

Axelrod, R., and W. D. Hamilton. 1981. The evolution of cooperation. *Science* 211:1390–1396.

Ayala, F. J. 1970. Teleological explanations in evolutionary biology. *Philosophy of Science* 37:1–15.

Bacolla, A., M. J. Ulrich, J. E. Larson, T. J. Ley, and R. D. Wells. 1995. An intramolecular triplex in the human γ-globin 5′-flanking region is altered by point mutations associated with hereditary persistence of fetal hemoglobin. *Journal of Biological Chemistry* 270:24556–24563.

Bacon, F. 1596. *Maxims of the Law*. London.

Bacon, F. 1605/1885. *The Advancement of Learning*. Ed. W. A. Wright. Oxford: Clarendon Press.

Bacon, F. 1623/1829. *De dignitate et augmentis scientiarum*. Ed. P. Mayer. Nuremberg: Riegell and Wiessner.

Badcock, C., and B. Crespi. 2006. Imbalanced genomic imprinting in brain development: An evolutionary basis for the aetiology of autism. *Journal of Evolutionary Biology* 19:1007–1032.

Baldwin, J. M. 1896. A new factor in evolution. *American Naturalist* 30:441–451, 536–553.

Bamford, D. H. 2003. Do viruses form lineages across different domains of life? *Research in Microbiology* 154:231–236.

Bamford, D. H., J. M. Grimes, and D. I. Stuart. 2008. What does structure tell us about virus evolution? *Current Opinion in Structural Biology* 15:655–663.

Barrett, P. H., P. J. Gautrey, S. Herbert, D. Kohn, and S. Smith. 1987. *Charles Darwin's Notebooks, 1836–1844*. Ithaca, NY: Cornell University Press.

Barton, N. H. 1995. A general model for the evolution of recombination. *Genetical Research* 65:123–144.

Bateson, G. 1972. *Steps to an Ecology of Mind*. Chicago: University of Chicago Press.

Beatty, J. 1994. The proximate/ultimate distinction in the multiple careers of Ernst Mayr. *Biology & Philosophy* 9:333–356.

Bendor, J., and P. Swistak. 1997. The evolutionary stability of cooperation. *American Political Science Review* 91:290–307.

Benirschke, K., J. M. Anderson, and L. E. Brownhill. 1962. Marrow chimerism in marmosets. *Science* 138:513–515.

Benne, R. 1994. RNA editing in trypanosomes. *European Journal of Biochemistry* 221:9–23.

Benson, S. D., J. K. H. Bamford, D. H. Bamford, and R. M. Burnett. 2004. Does common architecture reveal a viral lineage spanning all three domains of life? *Molecular Cell* 16:673–685.

Bergstrom, C. T., and M. Rosvall. 2011. The transmission sense of information. *Biology & Philosophy* 26:159–176.

Berkeley, M. J. 1857. *Introduction to Cryptogamic Botany*. London: Bailliere.

Bestor, T. H. 1990. DNA methylation: Evolution of a bacterial immune function into a regulator of gene expression and genome structure in higher eukaryotes. *Philosophical Transactions of the Royal Society B* 326:179–187.

Bestor, T. H., and B. Tycko. 1996. Creation of genomic methylation patterns. *Nature Genetics* 12:363–367.

Bianchi, D. W., G. K. Zickwolf, G. J. Weil, S. Sylvester, and M. A. DeMaria. 1996. Male fetal progenitor cells persist in maternal blood for as long as 27 years postpartum. *Proceedings of the National Academy of Sciences USA* 93:705–708.

Bird, A. P. 1993. Functions for DNA methylation in vertebrates. *Cold Spring Harbor Symposia on Quantitative Biology* 58:281–285.

Bird, A. P. 1995. Gene number, noise reduction and biological complexity. *Trends in Genetics* 11:94–100.

Bock, W. J., and G. Wahlert. 1965. Adaptation and the form–function complex. *Evolution* 19:269–299.

Boghardt, T. 2012. *The Zimmermann Telegram*. Annapolis: Naval Institute Press.

Boke, N. H. 1940. Histology and morphogenesis of the phyllode in certain species of *Acacia*. *American Journal of Botany* 27:73–90.

Borges, J. L. 2000. The Total Library (1939). Trans. E. Weinberger. In *The Total Library: Non-Fiction 1922–1986*, ed. E. Weinberger, 214–216. London: Allen Lane.

Bouchard, F., and A. Rosenberg. 2004. Fitness, probability and the principles of natural selection. *British Journal for the Philosophy of Science* 55:693–712.

Bouillard, F., D. Ricquier, G. Mory, and J. Thibault. 1984. Increased level of mRNA for the uncoupling protein in brown adipose tissue of rats during thermogenesis induced by cold exposure or norepinephrine infusion. *Journal of Biological Chemistry* 259:11583–11586.

Bowler, P. J. 1983. *The Eclipse of Darwinism*. Baltimore: Johns Hopkins University Press.

Bowler, P. J. 1992. *The Non-Darwinian Revolution*. Baltimore: Johns Hopkins University Press.

Boyd, R., P. J. Richerson, and J. Henrich. 2011. The cultural niche: Why social learning is essential for human adaptation. *Proceedings of the National Academy of Sciences USA* 108:10918–10925.

Breaker, R. R. 2008. Complex riboswitches. *Science* 319:1795–1797.

Breaker, R. R. 2012. Riboswitches and the RNA world. *Cold Spring Harbor Perspectives in Biology* 4:a003566.

Brion, P., and E. Westhof. 1997. Hierarchy and dynamics of RNA folding. *Annual Review of Biophysics and Biomolecular Structure* 26: 113–137.

Brown, V. 1994. *Adam Smith's Discourse*. London: Routledge.

Brown, W. M., and N. S. Consedine. 2004. Just how happy is the happy puppet? An emotional signaling and kinship theory perspective on the behavioral phenotype of Angelman syndrome children. *Medical Hypotheses* 63:377–385.

Brunekreef, G. A., H. J. Kraft, J. G. G. Schoenmakers, and N. H. Lubsen. 1996. Mechanism of recruitment of the lactate dehydrogenase-B/ε-crystallin gene by the duck lens. *Journal of Molecular Biology* 262:629–639.

Burnham, T. C., and D. D. P. Johnson. 2005. The biological and evolutionary logic of human cooperation. *Analyse Kritik* 27:113–135.

Buss, L. W. 1982. Somatic cell parasitism and the evolution of somatic tissue compatibility. *Proceedings of the National Academy of Sciences USA* 79:5337–5341.

Butler, M. G. 1990. Prader-Willi syndrome: Current understanding of cause and diagnosis. *American Journal of Medical Genetics* 35:319–332.

Calcott, B. 2014. Engineering and evolvability. *Biology & Philosophy* 29:293–313.

Cannon, B., and J. Nedergaard. 2004. Brown adipose tissue: Function and physiological significance. *Physiological Reviews* 84:277–359.

Cannon, W. B. 1945. *The Way of an Investigator*. New York: W. W. Norton.

Capurro, R., and B. Hjørland. 2003. The concept of information. *Annual Review of Information Science and Technology* 37:343–411.

Cartmill, M. 1994. A critique of homology as a morphological concept. *American Journal of Physical Anthropology* 94:115–123.

Cassidy, S. B. 1988. Management of the problems of infancy: Hypotonia, developmental delay, and feeding problems. In *Prader-Willi Syndrome: Selected Research and Management Issues*, ed. M. L. Caldwell and R. L. Taylor, 43–51. New York: Springer.

Chapman, J. 1865. *Diarrhœa and Cholera: Their Origin, Proximate Cause, and Cure, through the Agency of the Nervous System, by Means of Ice*. London: Trübner.

Charlesworth, B. 1990. Mutation-selection balance and the evolutionary advantage of sex and recombination. *Genetical Research* 55:199–221.

Chaudhry, A., R. G. MacKenzie, L. M. Georgic, and J. G. Granneman. 1994. Differential interaction of β_1- and β_3-adrenergic receptors with G_i in rat adipocytes. *Cellular Signaling* 6:457–465.

Chuong, E. B. 2013. Retroviruses facilitate the rapid evolution of the mammalian placenta. *BioEssays* 35:853–861.

Chuong, E. B., M. A. K. Rumi, M. J. Soares, and J. C. Baker. 2013. Endogenous retroviruses function as species-specific enhancer elements in the placenta. *Nature Genetics* 45:325–329.

Clarke, D. J., and G. Marston. 2000. Problem behaviors associated with 15q–Angelman syndrome. *American Journal on Mental Retardation* 105:25–31.

Clayton-Smith, J. 1993. Clinical research on Angelman syndrome in the United Kingdom: Observations on 82 affected kindreds. *American Journal of Medical Genetics* 46:12–15.

Coase, R. H. 1993. The nature of the firm (1937). In *The Nature of the Firm*, ed. O. E. Williamson and S. G. Winter, 18–33. New York: Oxford University Press.

Cohen, C. J., W. M. Lock, and D. L. Mager. 2009. Endogenous retroviral LTRs as promoters for human genes: A critical assessment. *Gene* 448:105–114.

Cohen, J. 2002. The immunopathogenesis of sepsis. *Nature* 420:885–891.

Colgate, S. A., and H. Ziock. 2011. A definition of information, the arrow of information, and its relationship to life. *Complexity* 16:54–62.

Collins, J. A., I. Irnov, S. Baker, and W. C. Winkler. 2007. Mechanism of mRNA destabilization by the *glmS* ribozyme. *Genes & Development* 21:3356–3368.

Cosmides, L. M., and J. Tooby. 1981. Cytoplasmic inheritance and intragenomic conflict. *Journal of Theoretical Biology* 89:83–129.

Cranefield, P. F. 1957. The organic physics of 1847 and the biophysics of today. *Journal of the History of Medicine and Allied Sciences* 12:407–423.

Csermely, P., R. Palotai, and R. Nussinov. 2010. Induced fit, conformational selection and independent dynamic segments: An extended view of binding events. *Trends in Biochemical Sciences* 35:539–546.

Cummins, R. 2002. Neo-teleology. In *Functions: New Essays in the Philosophy of Psychology and Biology*, ed. A. Ariew, R. Cummins, and M. Perlman, 157–172. Oxford: Oxford University Press.

Cuvier, G. 1817. *Le règne animal distribué d'après son organisation*, volume 1. Paris: Deterville.

Darwin, C. 1859. *On the Origin of Species by Means of Natural Selection, or The Preservation of Favoured Races in the Struggle for Life*. London: John Murray.

Darwin, C. 1862. *On the Various Contrivances by Which British and Foreign Orchids Are Fertilised by Insects, and on the Good Effects of Intercrossing*. London: John Murray.

Darwin, C. 1883/1998. *The Variation of Animals and Plants under Domestication*, volume 2. 2nd ed. Baltimore: Johns Hopkins University Press.

Darwin, E. 1818. *Zoonomia, or The Laws of Organic Life*. 4th American ed. Philadelphia: Edward Earle.

Darwin, F. 1898. *The Life and Letters of Charles Darwin*, volume 2. London: John Murray.

Davis, L. I. 1995. The nuclear pore complex. *Annual Review of Biochemistry* 64:865–896.

Dawkins, R. 1976. *The Selfish Gene*. Oxford: Oxford University Press.

Dawkins, R. 1982. *The Extended Phenotype*. Oxford: Oxford University Press.

Dawkins, R. 1989. The evolution of evolvability. In *Artificial Life*, ed. C. G. Langton, 201–220. Reading, MA: Addison-Wesley.

DeChiara, T. M., E. J. Robertson, and A. Efstratiadis. 1991. Parental imprinting of the mouse insulin-like growth factor II gene. *Cell* 64:849–859.

Delaine, C., C. L. Alvino, K. A. McNeil, T. D. Mulhern, L. Gauguin, P. De Meyts, et al. 2007. A novel binding site for the human insulin-like growth factor-II (IGF-II)/mannose 6-phosphate receptor on IGF-II. *Journal of Biological Chemistry* 282:1886–18894.

Delbrück, M. 1949. A physicist looks at biology. *Transactions of the Connecticut Academy of Arts and Sciences* 38:173–190.

Delbrück, M. 1971. Aristotle-totle-totle. In *Of Microbes and Life*, ed. J. Monod and E. Borek, 50–55. New York: Columbia University Press.

Dennett, D. C. 1984. *Elbow Room*. Cambridge, MA: MIT Press.

Dennett, D. C. 1987. *The Intentional Stance*. Cambridge, MA: MIT Press.

Dennett, D. C. 1991. *Consciousness Explained*. Boston: Little, Brown.

Dennett, D. C. 1995. *Darwin's Dangerous Idea*. New York: Simon and Schuster.

Dennett, D. C. 2005. *Sweet Dreams*. Cambridge, MA: MIT Press.

Dennett, D. C. 2009. Darwin's "strange inversion of reasoning." *Proceedings of the National Academy of Sciences USA* 106:10061–10065.

Dennett, D. C. 2011. Homunculi rule: Reflections on *Darwinian Populations and Natural Selection* by Peter Godfrey-Smith. *Biology & Philosophy* 26:475–488.

Dennett, D. C. 2014. The evolution of reasons. In *Contemporary Philosophical Naturalism and Its Implications*, ed. B. Bashour and H. D. Muller, 47–62. New York: Routledge.

Derrida, J. 1967. *De la grammatologie*. Paris: Éditions de Minuit.

Derrida, J. 1976. *Of Grammatology*. Trans. G. C. Spivak. Baltimore: Johns Hopkins University Press.

Derrida, J. 2016. *Of Grammatology*. 40th anniv. ed. Trans. G. C. Spivak. Baltimore: Johns Hopkins University Press.

Descartes, R. 1641/2011. *Meditations and Other Writings*. London: Folio Society.

Descartes, R. 1647/1983. *Principles of Philosophy*. Trans. V. R. Miller and R. P. Miller. Dordrecht: Reidel.

Dewey, J. 1896. The reflex arc concept in psychology. *Psychological Review* 3:357–370.

Dewsbury, D. A. 1999. The proximate and the ultimate: Past, present, and future. *Behavioral Processes* 46:189–199.

Dickens, C. 1849. *David Copperfield*. London: Bradbury & Evans.

Dickins, T. E., and R. A. Barton. 2013. Reciprocal causation and the proximate–ultimate distinction. *Biology & Philosophy* 28:747–756.

Didden, R., H. Korzilius, P. Duker, and L. M. G. Curfs. 2004. Communicative functioning in individuals with Angelman syndrome: A comparative study. *Disability and Rehabilitation* 26:1263–1267.

Dietrich, M. R. 1998. Paradox and persuasion: Negotiating the place of molecular evolution within evolutionary biology. *Journal of the History of Biology* 31:85–111.

Dill, K. A. 1999. Polymer principles and protein folding. *Protein Science* 8:1166–1180.

Dilthey, W. 1883. *Einleitung in die Geisteswissenschaften: Versuch einer Grundlegung für daß Studium der Gesellschaft und der Geschichte*. Leipzig: Dunder & Humblot.

Dilthey, W. 1894. Ideen über eine beschreibende und zergliedernde Psychologie. *Sitzungsberichte der Königlich Preußischen Akademie der Wissenschaften zu Berlin* 2 Hb: 1309–1407.

Dilthey, W. 1900. Die Entstehung der Hermeneutik. In *Philosophische Abhandlungen, Christoph Sigwart zu seinem siebigsten Geburtstage 28. März 1900*, 185–202. Tübingen: J. C. B. Mohr (Paul Siebeck).

Dilthey, W. 1979. Ideas about a descriptive and analytical psychology. Trans. of excerpts by H. P. Rickman. In *W. Dilthey: Selected Writings*, 88–97. Cambridge: Cambridge University Press.

Dilthey, W. 1989. *Introduction to the Human Sciences*. Princeton: Princeton University Press.

Dilthey, W. 1996. The rise of hermeneutics. Trans. F. R. Jameson and R. A. Makkreel. In *Hermeneutics and the Study of History: Wilhelm Dilthey, Selected Works*, volume 4, ed. R. A. Makkreel and F. Rodi, 235–258. Princeton: Princeton University Press.

Dover, G. A. 1993. Evolution of genetic redundancy for advanced players. *Current Opinion in Genetics and Development* 3:902–910.

du Bois-Reymond, E. 1918. *Jugendbriefe von Emil du Bois-Reymond an Eduard Hallmann*. Berlin: Reimer.

Duesterberg, V. K., I. T. Fischer-Hwang, C. F. Perez, D. W. Hogan, and S. M. Block. 2015. Observation of long-range tertiary interactions during ligand binding by the TPP riboswitch aptamer. *eLife* 4:e12362.

Duker, P., S. van Driel, and J. van de Bercken. 2002. Communication profiles of individuals with Down's syndrome, Angelman syndrome and pervasive developmental disorder. *Journal of Intellectual Disability Research* 46:35–40.

Dupré, J. 2017. The metaphysics of evolution. *Interface Focus* 7:20160148.

Eberhard, W. G. 1980. Evolutionary consequences of intracellular organelle competition. *Quarterly Review of Biology* 55:231–249.

Eckermann, J. P. 1836. *Gespräche mit Goethe in den lezten Jahren seines Lebens 1823–1832*. Zwenter Theil. Leipzig: Brodhaus.

Eimer, T. *On Orthogenesis and the Impotence of Natural Selection in Species-Formation*. Chicago: Open Court, 1898.

Ellington, A. D., and J. W. Szostak. 1990. *In vitro* selection of RNA molecules that bind specific ligands. *Nature* 346:818–822.

Emera, D., C. Casola, V. J. Lynch, D. E. Wildman, D. Agnew, and G. P. Wagner. 2012. Convergent evolution of endometrial prolactin expression in primates, mice, and elephants through the independent recruitment of transposable elements. *Molecular Biology and Evolution* 29:239–247.

Emera, D., and G. P. Wagner. 2012a. Transformation of a transposon into a derived prolactin promoter with function during human pregnancy. *Proceedings of the National Academy of Sciences USA* 109:11246–11251.

Emera, D., and G. P. Wagner. 2012b. Transposable element recruitments in the mammalian placenta: Impacts and mechanisms. *Briefings in Functional Genomics* 11:267–276.

Errington, J. 1996. Determination of cell fate in *Bacillus subtilis*. *Trends in Genetics* 12:31–34.

Eshel, I. 1985. Evolutionary genetic stability of Mendelian segregation and the role of free recombination in the chromosomal system. *American Naturalist* 125:412–420.

Eshel, I. 1996. On the changing concept of evolutionary population stability as a reflection of a changing point of view in the quantitative theory of evolution. *Journal of Mathematical Biology* 34:485–510.

Etzel, M., and M. Mörl. 2017. Synthetic riboswitches: From plug and pray toward plug and play. *Biochemistry* 56:1181–1198.

Ewens, W. J. 2011. What is the gene trying to do? *British Journal for the Philosophy of Science* 62:155–176.

Filson, A. J., A. Louvi, A. Efstratiadis, and E. J. Robertson. 1993. Rescue of the T-associated maternal effect in mice carrying null mutations in *Igf-2* and *Igf2r*, two reciprocally imprinted genes. *Development* 118:731–736.

Finnegan, D. J. 2012. Retrotransposons. *Current Biology* 22:R432–R437.

Fisher, R. A. 1934. Indeterminism and natural selection. *Philosophy of Science* 1:99–117.

Fisher, R. A. 1941. Average excess and average effect of a gene substitution. *Annals of Eugenics* 11:53–63.

Fisher, R. A. 1950. *Creative Aspects of Natural Law.* Cambridge: Cambridge University Press.

Fisher, R. A. 1958. *The Genetical Theory of Natural Selection.* 2nd ed. New York: Dover.

Fitch, W. M. 1970. Distinguishing homologous from analogous proteins. *Systematic Zoology* 19:99–113.

FitzGibbon, C. D., and J. H. Fanshawe. 1988. Stotting in Thomson's gazelles: An honest signal of condition. *Behavioral Ecology and Sociobiology* 23:69–74.

Fogassi, L. 2010. The mirror neuron system: How cognitive functions emerge from motor organization. *Journal of Economic Behavior and Organization* 77:66–75.

Fonstein, M., and R. Haselkorn. 1995. Physical mapping of bacterial genomes. *Journal of Bacteriology* 177:3361–3369.

Francis, R. C. 1990. Causes, proximate and ultimate. *Biology & Philosophy* 5:401–415.

Frank, S. A. 2009. Natural selection maximizes Fisher information. *Journal of Evolutionary Biology* 22:231–244.

Frank, S. A. 2012. Natural selection. III. Selection versus transmission and the levels of selection. *Journal of Evolutionary Biology* 25:227–243.

Frede, M. 1980. The original notion of cause. In *Doubt and Dogmatism*, ed. M. Schofield, M. Burnyeat, and J. Barnes, 217–249. Oxford: Clarendon Press.

Freeman, P. 2006. The Zimmermann telegram revisited: A reconciliation of the primary sources. *Cryptologia* 30:98–150.

Friedman, W. F., and C. J. Mendelsohn. 1994. *The Zimmermann Telegram of January 16, 1917 and Its Cryptographic Background*. Rev. ed. Laguna Hills, CA: Aegean Park Press.

Gadamer, H.-G. 1992. *Truth and Method*. 2nd. rev. edition. Trans. J. Weinsheimer and D. G. Marshall. New York: Crossroads.

Galantucci, B., C. A. Fowler, and M. T. Turvey. 2006. The motor theory of speech perception reviewed. *Psychonomic Bulletin & Review* 13:361–377.

Gallese, V. 2007. Before and below "theory of mind": Embodied simulation and the neural correlates of social cognition. *Philosophical Transactions of the Royal Society B* 362:659–669.

Gardner, A. 2013. Ultimate explanations concern the adaptive rationale for organism design. *Biology & Philosophy* 28:787–791.

Gentilucci, M., and M. C. Corballis. 2006. From manual gesture to speech: A gradual transition. *Neuroscience and Biobehavioral Reviews* 30:949–960.

Gerhart, J., and M. Kirschner. 2009. The theory of facilitated variation. *Proceedings of the National Academy of Sciences USA* 104:8582–8589.

Gibson, W., and L. Garside. 1990. Kinetoplast DNA minicircles are inherited from both parents in genetic hybrids of *Trypanosoma brucei*. *Molecular and Biochemical Parasitology* 42:45–54.

Gliddon, C. J., and P. H. Gouyon. 1989. The units of selection. *Trends in Ecology and Evolution* 4:204–208.

Godfrey-Smith, P. 2009. *Darwinian Populations and Natural Selection*. Oxford: Oxford University Press.

Goodey, N. M., and S. J. Benkovic. 2008. Allosteric regulation and catalysis emerge via a common route. *Nature Chemical Biology* 4:474–482.

Gornik, S. G., K. L. Ford, T. D. Mulhern, A. Bacic, G. I. McFadden, and R. F. Waller. 2012. Loss of nucleosomal DNA condensation coincides with appearance of a novel nuclear protein in dinoflagellates. *Current Biology* 22:2303–2312.

Gould, S. J. 1980. Sociobiology and the theory of natural selection. In *Sociobiology: Beyond Nature/Nurture? Reports, Definitions and Debate*, ed. G. W. Barlow and J. Silverberg, 257–269. Boulder, CO: Westview Press.

Gould, S. J. 1990. *Wonderful Life: The Burgess Shale and the Nature of History*. London: Hutchinson.

Gould, S. J., and R. C. Lewontin. 1979. The spandrels of San Marco and the Panglossian paradigm: A critique of the adaptationist programme. *Philosophical Transactions of the Royal Society B* 205:581–598.

Gould, S. J., and E. S. Vrba. 1982. Exaptation—a missing term in the science of form. *Paleobiology* 8:4–15.

Gouyon, P. H., and C. Gliddon. 1988. The genetics of information and the evolution of avatars. In *Population Genetics and Evolution*, ed. G. de Jong, 119–123. Berlin: Springer.

Grafen, A. 2006. Optimization of inclusive fitness. *Journal of Theoretical Biology* 238:541–563.

Grafen, A. 2014. The formal Darwinism project in outline. *Biology & Philosophy* 29:155–174.

Gray, A. 1874. Scientific worthies. III.—Charles Robert Darwin. *Nature* 10:79–81.

Gray, R. 1992. Death of the gene: Developmental systems strike back. In *Trees of Life: Essays in Philosophy of Biology*, ed. P. Griffiths, 165–209. Dordrecht: Kluwer.

Graw, J. 2009. Genetics of crystallins: Cataract and beyond. *Experimental Eye Research* 88:173–189.

Gregory, R. L. 1981. *Mind in Science*. Cambridge: Cambridge University Press.

Grene, M. 1972. Aristotle and modern biology. *Journal of the History of Ideas* 33:395–424.

Grieco, J. C., R. H. Bahr, M. R. Schoenberg, L. Conover, L. N. Mackie, and E. J. Weeber. 2018. Quantitative measurement of communication ability in children with Angelman syndrome. *Journal of Applied Research in Intellectual Disabilities* 31:e49–e58.

Griffiths, P. E. 1998. What is the developmentalist challenge? *Philosophy of Science* 65:253–258.

Griffiths, P. E. 2013. Lehrman's dictum: Information and explanation in developmental biology. *Developmental Psychobiology* 55:22–32.

Griffiths, P. E., and E. M. Neumann-Held. 1999. The many faces of the gene. *BioScience* 49:656–662.

Guerrero-Bosagna, C. 2012. Finalism in Darwinian and Lamarckian evolution: Lessons from epigenetics and developmental biology. *Evolutionary Biology* 39:283–300.

Hackett, J. A., R. Sengupta, J. J. Zylicz, K Murakami, Ca. Lee, T. A. Down, and M. A. Surani. 2013. Germline DNA demethylation dynamics and imprint erasure through 5-hydroxymethylcytosine. *Science* 339:448–452.

Haidt, J. 2001. The emotional dog and its rational tail: A social intuitionist approach to moral judgment. *Psychological Review* 108:814–834.

Haig, D. 1992a. Genomic imprinting and the theory of parent-offspring conflict. *Seminars in Developmental Biology* 3:153–160.

Haig, D. 1992b. Intragenomic conflict and the evolution of eusociality. *Journal of Theoretical Biology* 156:401–403.

Haig, D. 1993a. The evolution of unusual chromosomal systems in sciarid flies: Intragenomic conflict and the sex ratio. *Journal of Evolutionary Biology* 6:249–261.

Haig, D. 1993b. Genetic conflicts in human pregnancy. *Quarterly Review of Biology* 68:495–532.

Haig, D. 1996a. Gestational drive and the green-bearded placenta. *Proceedings of the National Academy of Sciences USA* 93:6547–6551.

Haig, D. 1996b. Placental hormones, genomic imprinting, and maternal–fetal communication. *Journal of Evolutionary Biology* 9:357–380.

Haig, D. 1997. Parental antagonism, relatedness asymmetries, and genomic imprinting. *Proceedings of the Royal Society B* 264:1657–1662.

Haig, D. 1999. Multiple paternity and genomic imprinting. *Genetics* 151:1229–1231.

Haig, D. 2000a. Genomic imprinting, sex-biased dispersal, and social behavior. *Annals of the New York Academy of Sciences* 907:149–163.

Haig, D. 2000b. The kinship theory of genomic imprinting. *Annual Review of Ecology and Systematics* 31:9–32.

Haig, D. 2002. *Genomic Imprinting and Kinship*. New Brunswick: Rutgers University Press.

Haig, D. 2003. Meditations on birthweight: Is it better to reduce the variance or increase the mean? *Epidemiology* 14:490–492 (erratum *Epidemiology* 14:632).

Haig, D. 2004a. Genomic imprinting and kinship: How good is the evidence? *Annual Review of Genetics* 38:553–585.

Haig, D. 2004b. The inexorable rise of gender and the decline of sex: Social change in academic titles, 1945–2001. *Archives of Sexual Behavior* 33:87–96.

Haig, D. 2006. Intragenomic politics. *Cytogenetic and Genome Research* 113:68–74.

Haig, D. 2007a. The amoral roots of morality. In *Biomedical Ethics*, ed. D. Steinberg, 25–28. Lebanon, NH: University Press of New England.

Haig, D. 2007b. Weismann rules! OK? Epigenetics and the Lamarckian temptation. *Biology & Philosophy* 22:415–428.

Haig, D. 2008a. Huddling: Brown fat, genomic imprinting, and the warm inner glow. *Current Biology* 18:R172–R174.

Haig, D. 2008b. Placental growth hormone-related proteins and prolactin-related proteins. *Placenta* 29 (supplement A): S36–S41.

Haig, D. 2010a. The huddler's dilemma: A cold shoulder or a warm inner glow. In *Social Behaviour: Genes, Ecology and Evolution*, ed. T. Székely, A. J. Moore, and J. Komdeur, 107–109. Cambridge: Cambridge University Press.

Haig, D. 2010b. Transfers and transitions: Parent-offspring conflict, genomic imprinting, and the evolution of human life history. *Proceedings of the National Academy of Sciences USA* 107:1731–1735.

Haig, D. 2011a. Does heritability hide in epistasis between linked SNPs? *European Journal of Human Genetics* 19:123.

Haig, D. 2011b. Genomic imprinting and the evolutionary psychology of human kinship. *Proceedings of the National Academy of Sciences USA* 108:10878–10885.

Haig, D. 2012. Retroviruses and the placenta. *Current Biology* 22:R609–R613.

Haig, D. 2013. Genomic vagabonds: Endogenous retroviruses and placental evolution. *BioEssays* 35:845–846.

Haig, D. 2014. Troubled sleep: Night waking, breastfeeding, and parent–offspring conflict. *Evolution, Medicine, and Public Health* 2014:32–39.

Haig, D. 2016. Transposable elements: Self-seekers of the germline, team players of the soma. *Bioessays* 38:1158–1166.

Haig, D. 2017. The extended reach of the selfish gene. *Evolutionary Anthropology* 26:95–97.

Haig, D., and A. Grafen. 1991. Genetic scrambling as a defence against meiotic drive. *Journal of Theoretical Biology* 153:531–558.

Haig, D., and C. Graham. 1991. Genomic imprinting and the strange case of the insulin-like growth factor-II receptor. *Cell* 64:1045–1046.

Haig, D., and S. Henikoff. 2004. Deciphering the genomic palimpsest. *Current Opinion in Genetics and Development* 14:599–602.

Haig, D., and R. Trivers. 1995. The evolution of parental imprinting: A review of hypotheses. In *Genomic Imprinting: Causes and Consequences*, ed. R. Ohlsson, K. Hall, and M. Ritzen, 17–28. Cambridge: Cambridge University Press.

Haig, D., and M. Westoby. 1989. Parent-specific gene expression and the triploid endosperm. *American Naturalist* 134:147–155.

Haig, D., and R. Wharton. 2003. Prader-Willi syndrome and the evolution of human childhood. *American Journal of Human Biology* 15:320–329.

Haig, D., and J. F. Wilkins. 2000. Genomic imprinting, sibling solidarity, and the logic of collective action. *Philosophical Transactions of the Royal Society B* 355:1593–1597.

Haldane, J. B. S. 1955. Population genetics. *New Biology* 18: 34–51.

Hall, N. 2004. Two concepts of causation. In *Causation and Counterfactuals*, ed. J. Collins, N. Hall, and L. Paul, 225–276. Cambridge, MA: MIT Press.

Hamburger, V. 1980. Embryology and the modern synthesis in evolutionary theory. In *The Evolutionary Synthesis: Perspectives on the*

Unification of Biology, ed. E. Mayr and W. B. Provine, 97–112. Cambridge, MA: Harvard University Press.

Hamilton, W. D. 1964. The genetical evolution of social behaviour. *Journal of Theoretical Biology* 7:1–52.

Hamilton, W. D. 1966. The moulding of senescence by natural selection. *Journal of Theoretical Biology* 12:12–45.

Hamilton, W. D. 1967. Extraordinary sex ratios. *Science* 156:477–488.

Hamilton, W. D. 1975. Gamblers since life began: Barnacles, aphids, elms. *Quarterly Review of Biology* 50:175–180.

Hamilton, W. D. 1979. Wingless and fighting males in fig wasps and other insects. In *Sexual Selection and Reproductive Competition in Insects*, ed. M. S. Blum and N. A. Blum, 167–220. New York: Academic Press.

Hamilton, W. D., R. Axelrod, and R. Tanese. 1990. Sexual reproduction as an adaptation to resist parasites (a review). *Proceedings of the National Academy of Sciences USA* 87:3566–3573.

Hamilton, W. J., D. J. Boyd, and H. W. Mossman. 1947. *Human Embryology*. Baltimore: Williams & Wilkins.

Hammerstein, P. 1996. Darwinian adaptation, population genetics and the streetcar theory of evolution. *Journal of Mathematical Biology* 34:511–532.

Hanchard, N., A. Elzein, C. Trafford, K. Rockett, M. Pinder, M. Jallow, et al. 2007. Classical sickle beta-globin haplotypes exhibit a high degree of long-rang haplotype similarity in African and Afro-Caribbean populations. *BMC Genetics* 8:52.

Hastings, I. M. 1992. Population genetic aspects of deleterious cytoplasmic genomes and their effect on the evolution of sexual reproduction. *Genetical Research* 59:215–225.

Hastings, I. M. 2000. Models of human genetic disease: How biased are the standard formulae? *Genetical Research* 75:107–114.

Heap, S. H., M. Hollis, B. Lyons, R. Sugden, and A. Weale. 1992. *The Theory of Choice*. Oxford: Blackwell.

Held, V. 2006. *The Ethics of Care*. Oxford: Oxford University Press.

Helmholtz, H. 1861. On the application of the law of the conservation of force to organic nature. *Proceedings of the Royal Institute of Great Britain* 3:347–357.

Hickey, D. A. 1982. Selfish DNA: A sexually transmitted nuclear parasite. *Genetics* 101:519–531.

Hicks, G. R., and N. V. Raikhel. 1995. Protein import into the nucleus: An integrated view. *Annual Review of Cell and Developmental Biology* 11:155–188.

Hitchcock, C. 2007. Prevention, preemption, and the principle of sufficient reason. *Philosophical Review* 116:495–532.

Hochman, A. 2013. The phylogeny fallacy and the ontogeny fallacy. *Biology & Philosophy* 28:593–612.

Hofstadter, D. R. 1979. *Gödel, Escher, Bach*. New York: Basic Books.

Hofstadter, D. R. 2007. *I Am a Strange Loop*. New York: Basic Books.

Holm, V. A., S. B. Cassidy, M. G. Butler, J. M. Hanchett, L. R. Greenswag, B. Y. Whitman, and F. Greenberg. 1993. Prader-Willi syndrome: Consensus diagnostic criteria. *Pediatrics* 91:398–402.

Horsler, K., and C. Oliver. 2006a. The behavioral phenotype of Angelman syndrome. *Journal of Intellectual Disability Research* 50:33–53.

Horsler, K., and C. Oliver. 2006b. Environmental influences on the behavioral phenotype of Angelman syndrome. *American Journal on Mental Retardation* 111:311–321.

Hughes-Schrader, S. 1948. Cytology of coccids (Coccoidea-Homoptera). *Advances in Genetics* 2:127–203.

Hull, D. L. 1978. A matter of individuality. *Philosophy of Science* 45:335–360.

Hurst, L. D., and W. D. Hamilton. 1992. Cytoplasmic fusion and the nature of sexes. *Proceedings of the Royal Society B* 247:189–194.

Hume, D. 1748/2004 An Enquiry Concerning Human Understanding. Mineola, NY: Dover.

Huxley, T. H. 1859. Darwin on the origin of species. *Times* (London), December 26, 8–9.

Huxley, T. H. 1869. Anniversary address of the president. *Quarterly Journal of the Geological Society of London* 25:xxxviii–liii.

Huxley, T. H. 1869a. The Natural History of Creation.—By Dr. Ernst Haeckel. [*Natürliche Schöpfungs-Geschichte.*—Von Dr. Ernst Haeckel, Professor an der Universität Jena.] Berlin 1868. First Notice. *Academy* 1: 13–14 (October 9, 1869).

Ingram, G. I. C. 1976. The history of haemophilia. *Journal of Clinical Pathology* 29:469–479.

Isles, A. R., W. Davies, and L. S. Wilkinson. 2006. Genomic imprinting and the social brain. *Philosophical Transactions of the Royal Society B* 361:2229–2237.

Jacob, F. 1977. Evolution and tinkering. *Science* 196:1161–1166.

James, W. 1887. What is an instinct? *Scribner's Magazine* 1:355–365.

James, W. 1890/1983. *The Principles of Psychology*. Cambridge, MA: Harvard University Press.

Johannsen, W. 1909. *Elemente der exacten Erblichkeitslehre*. Jena: Gustav Fischer.

Johannsen, W. 1911. The genotype conception of heredity. *American Naturalist* 45:129–159.

Jolleff, N., and M. M. Ryan. 1993. Communication development in Angelman's syndrome. *Archives of Disease in Childhood* 69:148–150.

Jonas, H. 1966. *The Phenomenon of Life: Toward a Philosophical Biology*. New York: Harper & Rowe.

Joyce, G. F. 2002. The antiquity of RNA-based evolution. *Nature* 418:214–221.

Jurgenson, C. T., T. P. Begley, and S. E. Ealick. 2009. The structural and biochemical foundations of thiamin biosynthesis. *Annual Review of Biochemistry* 78:569–603.

Kaiser, D. 1986. Control of multicellular development: *Dictyostelium* and *Myxococcus*. *Annual Review of Genetics* 20:539–566.

Kant, I. 1781/1998. *Critique of Pure Reason*. Trans. and ed. P. Guyer and A. W. Wood. Cambridge: Cambridge University Press.

Kant, I. 1790/2000. *Critique of the Power of Judgment*. Trans. P. Guyer and E. Matthews. Cambridge: Cambridge University Press.

Kashtan, N., and U. Alon. 2005. Spontaneous evolution of modularity and network motifs. *Proceedings of the National Academy of Sciences USA* 102:13733–13778.

Keverne, E. B., R. Fundele, M. Narasimha, S. C. Barton, and M. A. Surani. 1996. Genomic imprinting and the differential roles of parental genomes in brain development. *Developmental Brain Research* 92:91–100.

Killian, J. K., C. M. Nolan, A. A. Wylie, T. Li, T. H. Vu, A. R. Hoffman, and R. L. Jirtle. 2001. Divergent evolution in *M6P/IGF2R* imprinting from the Jurassic to the Quaternary. *Human Molecular Genetics* 10:1721–1728.

Kingsley, C. 1873. *Madam How or Lady Why, or First Lessons in Earth Lore for Children*. 3rd. ed. London: Strahan.

Klein, D. J., and A. R. Ferré-D'Amaré. 2006. Structural basis of *glmS* ribozyme activation by glucosamine-6-phosphate. *Science* 313:1752–1756.

Knight, R., and M. Yarus. 2003. Finding specific RNA motifs: Function in a zeptomole world? *RNA* 9:218–230.

Kotler, J., and D. Haig. 2018. The tempo of human childhood: A maternal foot on the accelerator, a paternal foot on the break. *Evolutionary Anthropology* 27:80–91.

Kristeller, P. O. 1978. Humanism. *Minerva* 16:586–595.

Kusano, K., T. Naito, N. Handa, and I. Kobayashi. 1995. Restriction–modification systems as genomic parasites in competition for specific sequences. *Proceedings of the National Academy of Sciences USA* 92:11095–11099.

Kuwajima, T., I. Nishimura, and K. Yoshikawa. 2006. Necdin promotes GABAergic neuron differentiation in cooperation with Dlx homeodomain proteins. *Journal of Neuroscience* 26:5383–5392.

Laland, K. N., J. Odling-Smee, W. Hoppitt, and T. Uller. 2013a. More of how and why: Cause and effect in biology revisited. *Biology & Philosophy* 28:719–745.

Laland, K. N., J. Odling-Smee, W. Hoppitt, and T. Uller. 2013b. More of how and why: A response to commentaries. *Biology & Philosophy* 28:793–810.

Lankester, E. R. 1870. On the use of the term homology in modern zoology and the distinction between homogenetic and homoplastic agreements. *Annals and Magazine of Natural History* 6:34–43.

Lankester, E. R. 1876. Perigenesis v. pangenesis—Haeckel's new theory of heredity. *Nature* 14:235–238.

Lankester, E. R. 1890. *The Advancement of Science: Occasional Essays and Addresses*. London: MacMillan.

Law, R., and V. Hutson. 1992. Intracellular symbionts and the evolution of uniparental cytoplasmic inheritance. *Proceedings of the Royal Society B* 248:69–77.

Lau, M. M. H., C. E. H. Stewart, Z. Liu, H. Bhatt, P. Rotwein, and C. L. Stewart. 1994. Loss of the imprinted IGF2/cation-independent mannose 6-phosphate receptor results in fetal overgrowth and perinatal lethality. *Genes & Development* 8:2953–2963.

Lau, M. W. L., and A. R. Ferré-D'Amaré. 2013. An *in vitro* evolved *glmS* ribozyme has the wildtype fold but loses coenzyme dependence. *Nature Chemical Biology* 9:805–810.

Lazcano, A., R. Guerrero, L. Margulis, and J. Oró. 1988. The evolutionary transition from RNA to DNA in early cells. *Journal of Molecular Evolution* 27:283–290.

Le Guyader, H. 2004. *Geoffroy Saint-Hilaire: A Visionary Naturalist.* Trans. M. Grene. Chicago: University of Chicago Press.

Lehman, N., C. D. Arenas, W. A. White, and F. J. Schmidt. 2011. Complexity through recombination: From chemistry to biology. *Entropy* 13:17–37.

Lehnherr, H., E. Maguin, S. Jafri, and M. B. Yarmolinsky. 1993. Plasmid addiction genes of bacteriophage P1: *doc*, which causes cell death on curing of prophage, and *phd*, which prevents host death when prophage is retained. *Journal of Molecular Biology* 233:414–428.

Leigh, E. G. 1971. *Adaptation and Diversity*. San Francisco: Freeman Cooper.

Lenski, R. E., S. C. Simpson, and T. T. Nguyen. 1994. Genetic analysis of a plasmid-encoded, host genotype-specific enhancement of bacterial fitness. *Journal of Bacteriology* 176:3140–3147.

Leopold, P. E., M. Montal, and J. N. Onuchic. 1992. Protein folding funnels: A kinetic approach to the sequence–structure relationship. *Proceedings of the National Academy of Sciences USA* 89:8721–8725.

Lerner, D., ed. 1965. *Cause and Effect*. New York: Free Press.

Lewis, D. 1973. Causation. *Journal of Philosophy* 70:556–567.

Lewis, D. 2000. Causation as influence. *Journal of Philosophy* 97:182–197.

Lewontin, R. C. 1974. The analysis of variance and the analysis of causes. *American Journal of Human Genetics* 26:400–411.

Lewontin, R. C. 2000. Foreword. In S. Oyama, *The Ontogeny of Information*, 2nd ed., vii–xv. Durham, NC: Duke University Press.

Li, S., and R. R. Breaker. 2013. Eukaryotic TPP riboswitch regulation of alternative splicing involving long-distance base pairing. *Nucleic Acids Research* 41:3022–3031.

Lickliter, R., and T. D. Berry. 1990. The phylogeny fallacy: Developmental psychology's misapplication of evolutionary theory. *Developmental Review* 10:348–364.

Lindley, D. V. 2000. The philosophy of statistics. *Statistician* 49:293–337.

Lloyd, E. 2005. Why the gene will not return. *Philosophy of Science* 72:287–310.

Lloyd, S. 2009. A quantum of natural selection. *Nature Physics* 5:164–166.

Lorenz, M. G., and W. Wackerknagel. 1994. Bacterial gene transfer by natural genetic transformation in the environment. *Microbiological Reviews* 58:563–602.

Lynch, M. 2007. The evolution of genetic networks by non-adaptive processes. *Nature Reviews Genetics* 8:803–813.

Lynch, V. J., A. Tanzer, Y. Wang, F. C. Leung, B. Gellersen, D. Emera, and G. P. Wagner. 2008. Adaptive changes in the transcription factor HoxA-11 are essential for the evolution of pregnancy in mammals. *Proceedings of the National Academy of Sciences USA* 105:14928–14933.

Lynch, V. J., R. D. LeClerc, G. May, and G. P. Wagner. 2011. Transposon-mediated rewiring of gene regulatory networks contributed to the evolution of pregnancy in mammals. *Nature Genetics* 43:1154–1159.

Mantovani, G., S. Bondioni, M. Locatelli, C. Pedroni, A. G. Lania, E. Ferrantee, et al. 2004. Biallelic expression of the Gsα gene in human bone and adipose tissue. *Journal of Clinical Endocrinology and Metabolism* 89:6316–6319.

Maynard Smith, J. 1982. *Evolution and the Theory of Games.* Cambridge: Cambridge University Press.

Maynard Smith, J. 1987. How to model evolution. In *The Latest on the Best: Essays on Evolution and Optimality*, ed. J. Dupré, 119–131. Cambridge, MA: MIT Press.

Maynard Smith, J., and E. Szathmáry. 1993. The origin of chromosomes I. Selection for linkage. *Journal of Theoretical Biology* 164:437–446.

Maynard Smith, J., and E. Szathmáry. 1995. *The Major Transitions in Evolution*. Oxford: Oxford University Press.

Mayr, E. 1961. Cause and effect in biology. *Science* 134:1501–1506.

Mayr, E. 1965. Cause and effect in biology. In *Cause and Effect*, ed. D. Lerner, 33–50. New York: Free Press.

Mayr, E. 1969. Footnotes on the philosophy of biology. *Philosophy of Science* 36:197–202.

Mayr, E. 1974. Teleological and teleonomic, a new analysis. *Boston Studies in the Philosophy of Science* 14:91–117.

Mayr, E. 1992. The idea of teleology. *Journal of the History of Ideas* 53:117–135.

Mayr, E. 1993. Proximate and ultimate causations. *Biology & Philosophy* 8:93–94.

Mayr, E. 2004. *What Makes Biology Unique?* Cambridge: Cambridge University Press.

McCosh, J., and G. Dickie. 1856. *Typical Forms and Special Ends in Creation*. Edinburgh: Thomas Constable.

Mellor, J. 2005. The dynamics of chromatin remodeling at promoters. *Molecular Cell* 19:147–157.

Miller, F. D. 1999. Aristotle's philosophy of soul. *Review of Metaphysics* 53:309–337.

Miller, S. P., P. Riley, and M. I. Shevell. 1999. The neonatal presentation of Prader-Willi syndrome revisited. *Journal of Pediatrics* 134:226–228.

Millikan, R. G. 1989. In defense of proper functions. *Philosophy of Science* 56:288–302.

Millikan, R. G. 1999. Historical kinds and the "special sciences." *Philosophical Studies* 95:45–65.

Milner, K. M., E. E. Craig, R. J. Thompson, M. W. M. Veltman, N. S. Thomas, S. Roberts, et al. 2005. Prader-Willi syndrome: Intellectual abilities and behavioural features by genetic subtype. *Journal of Child Psychology and Psychiatry* 46:1089–1096.

Moffatt, B. 2011. Conflations in the causal account of information undermine the parity thesis. *Philosophy of Science* 78:284–302.

Molnar-Szakacs, I. 2010. From actions to empathy and morality—a neural perspective. *Journal of Economic Behavior and Organization* 77:76–85.

Monod, J., and F. Jacob. 1961. Teleonomic mechanisms in cellular metabolism, growth, and differentiation. *Cold Spring Harbor Symposia on Quantitative Biology* 26:389–401.

Monod, J., J. P. Changeux, and F. Jacob. 1963. Allosteric proteins and cellular control systems. *Journal of Molecular Biology* 6:306–329.

Montange, R. K., and P. T. Batey. 2008. Riboswitches: Emerging themes in RNA structure and function. *Annual Review of Biophysics* 37:117–133.

Monteverde, D. R., L. Gómez-Consarnau, C. Suffridge, and S. A. Sañudo-Wilhelmy. 2017. Life's utilization of B vitamins on early Earth. *Geobiology* 15:3–18.

Moore, T., and D. Haig. 1991. Genomic imprinting in mammalian development: A parental tug-of-war. *Trends in Genetics* 7:45–49.

Morrison, S. F. 2004. Central pathways controlling brown adipose tissue thermogenesis. *News in Physiological Sciences* 19:67–74.

Morton, N. E. 1991. Parameters of the human genome. *Proceedings of the National Academy of Sciences USA* 88:7474–7464.

Mullahy, J. H. 1951. Evolution in the plant kingdom. *Bios* 22:20–25.

Muller, H. J. 1922. Variation due to change in the individual gene. *American Naturalist* 56:32–50.

Muller, H. J. 1929. The method of evolution. *Scientific Monthly* 29:481–505.

Muller, H. J. 1949. The Darwinian and modern conceptions of natural selection. *Proceedings of the American Philosophical Society* 93:459–470.

Nag, N., K. Peterson, K. Wyatt, S. Hess, S. Ray, J. Favor, et al. 2007. Endogenous retroviral insertion in *Cryge* in the mouse *No3* cataract mutant. *Genomics* 89:512–520.

Nakamura, K., and S. F. Morrison. 2007. Central efferent pathways mediating skin-cooling evoked sympathetic thermogenesis in brown adipose tissue. *American Journal of Physiology* 292:R127–R136.

Nash, O. 1936. *The Bad Parents' Book of Verse*. New York: Simon and Schuster.

Nauert, C. G. 1990. Humanist infiltration into the academic world: Some studies of northern universities. *Renaissance Quarterly* 43:799–812.

Neander, K. 1991. Functions as selected effects: The conceptual analyst's defense. *Philosophy of Science* 58:168–184.

Neher, R. A., and B. I. Shraiman. 2009. Competition between recombination and epistasis can cause a transition from allele to genotype selection. *Proceedings of the National Academy of Sciences USA* 106:6866–6871.

Neher, R. A., T. A. Kessinger, and B. I. Shraiman. 2013. Coalescence and genetic diversity in sexual populations under selection. *Proceedings of the National Academy of Sciences USA* 110:15836–15841.

Nietzsche, F. 1886/1989. *Jenseits von Gut und Böse (Beyond Good and Evil)*. Trans. W. Kaufmann. New York: Random House.

Nietzsche, F. 1887. *Zur Genealogie der Moral*. Leipzig: Naumann.

Nietzsche, F. 2007. *On the Genealogy of Morality*. Rev. student ed. Trans. C. Diethe. Cambridge: Cambridge University Press.

Nordström, K., and S. J. Austin. 1989. Mechanisms that contribute to the stable segregation of plasmids. *Annual Review of Genetics* 23:37–69.

Nose, A., A. Nagafuchi, and M. Takeichi. 1988. Expressed recombinant cadherins mediate cell sorting in model systems. *Cell* 54:993–1001.

Nowak, M. A., and K. Sigmund. 1998. Evolution of indirect reciprocity by image scoring. *Nature* 393:573–577.

O'Day, D. H. 1979. Aggregation during sexual development in *Dictyostelium discoideum*. *Canadian Journal of Microbiology* 25:1416–1426.

Õiglane-Shlik, E., R. Zordania, H. Varendi, A. Antson, M.-L. Mägi, G. Tasa, et al. 2006. The neonatal phenotype of Prader-Willi syndrome. *American Journal of Medical Genetics* 140A:1241–1244.

Okada, T., O. P. Ernst, K. Palczewski, and K. P. Hofmann. 2001. Activation of rhodopsin: New insights from structural and biochemical studies. *Trends in Biochemical Sciences* 26:318–324.

Okasha, S. 2008. Fisher's fundamental theorem of natural selection—a philosophical analysis. *British Journal for the Philosophy of Science* 59:319–351.

Okasha, S. 2012. Social justice, genomic justice and the veil of ignorance: Harsanyi meets Mendel. *Economics and Philosophy* 28:43–71.

Oliver, C., L. Demetriades, and S. Hall. 2002. Effects of environmental events on smiling and laughing behavior in Angelman syndrome. *American Journal on Mental Retardation* 107:194–200.

Oliver, C., K. Horsler, K. Berg, G. Bellamy, K. Dick, and E. Griffiths. 2007. Genomic imprinting and the expression of affect in Angelman syndrome: What's in the smile? *Journal of Child Psychology and Psychiatry* 48:571–579.

Owen, R. 1846. Observations on Mr. Strickland's article on the structural relations of organized beings. *London, Edinburgh and Dublin Philosophical Magazine and Journal of Science* (third series) 28:525–527.

Owen, R. 1848. *On the Archetype and Homologies of the Vertebrate Skeleton*. London: van Voorst.

Owen, R. 1849. *On the Nature of Limbs*. London: Van Voorst.

Owen, R. 1868. *On the Anatomy of Vertebrates*, volume 3: *Mammals*. London: Longmans Green.

Oyama, S. 2000. *The Ontogeny of Information*. 2nd ed. Durham, NC: Duke University Press.

Pancer, Z., H. Gershon, and B. Rinkevich. 1995. Coexistence and possible parasitism of somatic and germ cell lines in chimeras of the colonial urochordate *Botryllus schlosseri*. *Biological Bulletin* 189:106–112.

Parnas, D. L. 1972. On the criteria to be used in decomposing systems into modules. *Communications ACM* 15:1053–1058.

Papineau, D. 1984. Representation and explanation. *Philosophy of Science* 51:550–572.

Pearl, J. 2000. *Causality*. Cambridge: Cambridge University Press.

Pearson, L. 1952. *Prophasis* and *aitia*. *Transactions and Proceedings of the American Philological Association* 83:203–223.

Peirce, C. S. 1877. Illustrations of the logic of science. First paper.—The fixation of belief. *Popular Science Monthly* 12:1–15.

Peirce, C. S. 1905. What pragmaticism is. *Monist* 15:161–181 [title corrected].

Pembrey, M. 1996. Imprinting and transgenerational modulation of gene expression; human growth as a model. *Acta Geneticae Medicae et Gemellologiae* 45:111–125.

Penner, K. A., J. Johnston, B. H. Faircloth, P. Irish, and C. A. Williams. 1993. Communication, cognition, and social interaction in the Angelman syndrome. *American Journal of Medical Genetics* 46:34–39.

Piatigorsky, J. 2007. *Gene Sharing and Evolution*. Cambridge, MA: Harvard University Press.

Pittendrigh, C. S. 1961. On temporal organization in living systems. *Harvey Lectures* 56:93–125.

Pittendrigh, C. S. 1993. Temporal organization: Reflections of a Darwinian clock-watcher. *Annual Review of Physiology* 55:17–54.

Plagge, A., E. Gordon, W. Dean, R. Boiani, S. Cinti, J. Peters, and G. Kelsey. 2004. The imprinted signaling protein XLαs is required for postnatal adaptation to feeding. *Nature Genetics* 36:818–826.

Plotkin, S. S., and J. N. Onuchic 2002. Understanding protein folding with energy landscape theory. Part 1: Basic considerations. *Quarterly Reviews of Biophysics* 35:111–116.

Popper, K. R. 2014. A new interpretation of Darwinism (1986). The first Medawar lecture. In *Karl Popper and the Two New Secrets of Life*, ed. H. J. Niemann, 115–129. Tübingen: Mohr Siebeck.

The potato disease. II. 1872. Editorial. *Nature* 6:409–410.

Poulton, E. B. 1908. *Essays on Evolution 1889–1907*. Oxford: Clarendon Press.

Price, G. R. 1995. The nature of selection. *Journal of Theoretical Biology* 175:389–396.

Punnett, R. C. 1913. *Mendelism*. 3rd ed. New York: MacMillan.

Queller, D. C. 2011. A gene's eye view of Darwinian populations. *Biology & Philosophy* 26:905–913.

Quine, W. V. 1987. *Quiddities*. Cambridge, MA: Harvard University Press.

Rawls, J. 1971. *A Theory of Justice*. Cambridge, MA: Harvard University Press.

Reddy, V. 2008. *How Infants Know Minds*. Cambridge, MA: Harvard University Press.

Redfield, R. J. 1993. Genes for breakfast: The have-your-cake-and-eat-it-too of bacterial transformation. *Journal of Heredity* 84:400–404.

Rice, W. R. 1987. The accumulation of sexually antagonistic genes as a selective agent promoting the evolution of reduced recombination between primitive sex chromosomes. *Evolution* 41:911–914.

Richerson, P. J., and R. Boyd. 2005. *Not by Genes Alone*. Chicago: University of Chicago Press.

Ricoeur, P. 1971. The model of the text: Meaningful action considered as a text. *Social Research* 38:529–562.

Ricoeur, P. 2004. *Memory, History, Forgetting*. Trans. K. Blamey and D. Pellauer. Chicago: Chicago University Press.

Ridge, K. D., N. G. Abulaev, M. Sousa, and K. Palczewski. 2003. Photo transduction: Crystal clear. *Trends in Biochemical Sciences* 28:479–487.

Ridley, M. 2000. *Mendel's Demon*. London: Weidenfeld and Nicholson.

Ridley, M., and A. Grafen. 1981. Are green beard genes outlaws? *Animal Behaviour* 29:954–955.

Rivier, D. H., and L. Pillus. 1994. Silencing speaks up. *Cell* 76:963–966.

Robins, S. J., J. M. Fasulo, G. M. Patton, E. J. Schaefer, D. E. Smith, and J. M. Ordovas. 1995. Gender differences in the development of hyperlipemia and atherosclerosis in hybrid hamsters. *Metabolism* 44:1326–1331.

Romanes, G. J. 1895. *Darwin, and after Darwin*. II. *Post-Darwinian Questions. Heredity and Utility*. Chicago: Open Court.

Romanovsky, A. A. 2007. Thermoregulation: Some concepts have changed. Functional architecture of the thermoregulatory system. *American Journal of Physiology* 292:R37–R46.

Roth, A., and R. R. Breaker. 2009. The structural and functional diversity of metabolite-binding riboswitches. *Annual Review of Biochemistry* 78:305–334.

Rowley, A., S. J. Dowell, and J. F. X. Diffley. 1994. Recent developments in the initiation of chromosomal DNA replication: A complex picture emerges. *Biochimica et Biophysica Acta* 1217:239–256.

Royer, M. 1975. Hermaphroditism in insects: Studies on *Icerya purchasi*. In *Intersexuality in the Animal Kingdom*, ed. R. Reinboth, 135–145. Berlin: Springer.

Rudolf Ludwig Karl Virchow: Obituary. 1902. *Nature* 66:551–552.

Russell, B. 1913. On the notion of cause. *Proceedings of the Aristotelian Society* 13:1–26.

Salmon, M. A., L. van Melderen, P. Bernard, and M. Couturier. 1994. The antidote and autoregulatory functions of the F plasmid CcdA protein: A genetic and biochemical survey. *Molecular and General Genetics* 244:530–538.

Sandhu, K. S., C. Shi, M. Sjölinder, Z. Zhao, A. Göndör, L. Liu, et al. 2009. Nonallelic transvection of multiple imprinted loci is organized by the *H19* imprinting control region during germline development. *Genes & Development* 23:2598–2603.

Sartorio, C. 2005. Causes as difference makers. *Philosophical Studies* 123:71–96.

Saumitou-Laprade, P., J. Cuguen, and P. Vernet. 1994. Cytoplasmic male sterility in plants: Molecular evidence and the nucleocytoplasmic conflict. *Trends in Ecology and Evolution* 9:431–435.

Saussure, F. de. 1916. *Cours de Linguistique Générale*. Paris: Payot.

Saussure, F. de. 1986. *Course in General Linguistics*. Trans. R Harris. La Salle, IL: Open Court.

Scantlebury, M., A. F. Russell, G. M. McIlraith, J. R. Speakman, and T. H. Clutton-Brock. 2002. The energetics of lactation in cooperatively breeding meerkats *Suricata suricatta*. *Proceedings of the Royal Society B* 269:2147–2153.

Selker, E. U. 1990. Premeiotic instability of repeated sequences in *Neurospora crassa*. *Annual Review of Genetics* 24:579–613.

Shannon, C. 1948. A mathematical theory of communication. *Bell System Technical Journal* 27:379–423, 623–656.

Shea, N. 2007. Representation in the genome and in other inheritance systems. *Biology & Philosophy* 22:313–331.

Shimkets, L. J. 1990. Social and developmental biology of the myxobacteria. *Microbiological Reviews* 54:473–501.

Smith, A. 1976. *The Theory of Moral Sentiments*. Oxford: Oxford University Press.

Sniegowski, P. D., and H. A. Murphy. 2006. Evolvability. *Current Biology* 16:R831–R834.

Sober, E. 1984. *The Nature of Selection*. Cambridge, MA: MIT Press.

Sober, E., and R. C. Lewontin. 1982. Artifact, cause, and genic selection. *Philosophy of Science* 49:157–180.

Sober, E., and D. S. Wilson. 1994. A critical review of philosophical work on the units of selection problem. *Philosophy of Science* 61:534–555.

Sober, E., and D. S. Wilson. 1998. *Unto Others*. Cambridge, MA: Harvard University Press.

Spencer, H. 1852. A theory of population deduced from the general law of animal fertility. *Westminster Review* (April):462–501.

Spencer, H. 1857. Progress: Its laws and cause. *Westminster Review* (April):445–485.

Spencer, H. 1867. *First Principles*. 2nd ed. London: Williams and Norgate.

Spinoza, B. 2002. *Spinoza: Complete Works*. Trans. S. Shirley. Ed. M. L. Morgan. Indianapolis: Hackett.

Sterne, L. 1767. *The Life and Opinions of Tristram Shandy, Gentleman*. York.

Sterelny, K., and P. E. Griffiths. 1999. *Sex and Death*. Chicago: Chicago University Press.

Sterelny, K., and P. Kitcher. 1988. The return of the gene. *Journal of Philosophy* 85:339–361.

Sterelny, K., K. C. Smith, and M. Dickison. 1996. The extended replicator. *Biology & Philosophy* 11:377–403.

Stewart, G. J., and C. A. Carlson. 1986. The biology of natural transformation. *Annual Review of Microbiology* 40:211–235.

Stock, D. W., J. M. Quattro, G. S. Whitt, and D. A. Powers. 1997. Lactate dehydrogenase (LDH) gene duplication during chordate evolution: The cDNA sequence of the LDH of the tunicate *Styela plicata*. *Molecular Biology and Evolution* 14:1273–1284.

Stoddard, C. D., R. K. Montange, S. P. Hennelly, R. P. Rambo, and K. Y. Sanbonmatsu. 2010. Free state conformational sampling of the SAM-I riboswitch aptamer domain. *Structure* 18:787–797.

Stokes, G. G. 1887. Science and revelation. *Nature* 36:333–335.

Strevens, M. 2013. Causality reunified. *Erkenntnis* 78:299–320.

Sturtevant, A. H. 1915. The behavior of the chromosomes as studied through linkage. *Zeitschrift für induktive Abstammungs- und Vererbungslehre* 13:234–287.

Su, T. T., P. J. Follette, and P. H. O'Farrell. 1995. Qualifying for the license to replicate. *Cell* 81:825–828.

Sudarsan, N., M. C. Hammond, K. F. Block, R. Welz, J. E. Barrick, A. Roth, and R. R. Breaker. 2006. Tandem riboswitch architectures exhibit complex gene control functions. *Science* 314:300–304.

Swainson, W. 1835. *A Treatise on the Geography and Classification of Animals*. London: Longman, Rees, Orme, Brown, Green, & Longman.

Terasawa, H., D. Kohda, H. Hatanaka, K. Nagata, N. Higashihashi, H. Fujiwara, et al. 1994. Solution structure of human insulin-like growth factor II; recognition sites for receptors and binding proteins. *EMBO Journal* 13:5590–5597.

Thierry. B. 2005. Integrating proximate and ultimate causation: Just one more go! *Current Science* 89:1180–1183.

Thisted, T., N. S. Sørensen, E. G. H. Wagner, and K. Gerdes. 1994. Mechanism of post-segregational killing: Sok antisense RNA interacts with Hok mRNA via its 5'-end single-stranded leader and competes with the 3'-end of Hok mRNA for binding to the *mok* translation initiation region. *EMBO Journal* 13:1960–1968.

Tinbergen, N. 1963. On aims and methods of ethology. *Zeitschrift für Tierpsychologie* 20:410–433.

Tinbergen, N. 1968. On war and peace in animals and man. *Science* 160:1411–1418.

Tononi, G. 2004. An information integration theory of consciousness. *BMC Neuroscience* 5:42.

Traulsen, A., and F. A. Reed. 2012. From genes to games: Cooperation and cyclic dominance in meiotic drive. *Journal of Theoretical Biology* 299:120–125.

Trivers, R. 1971. The evolution of reciprocal altruism. *Quarterly Review of Biology* 46:35–57.

Trivers, R. 1974. Parent-offspring conflict. *American Zoologist* 14:249–264.

Trivers, R. 2000. The elements of a scientific theory of self-deception. *Annals of the New York Academy of Sciences* 907:114–131.

Trivers, R. 2011. *Deceit and Self-Deception*. London: Allen Lane.

Tseng, Y. H., A. J. Butte, E. Kokkotou, V. K. Yechoor, C. M. Taniguchi, K. M. Kriauciunas, et al. 2005. Prediction of preadipocyte differentiation by gene expression reveals role of insulin receptor substrates and necdin. *Nature Cell Biology* 7:601–611.

van Dijk, B. A., D. I. Boomsma, and A. J. M. de Man. 1996. Blood group chimerism in multiple human births is not rare. *American Journal of Medical Genetics* 61:264–268.

van Rheede, T., R. Amons, N. Stewart, and W. W. de Jong. 2003. Lactate dehydrogenase A as a highly abundant eye lens protein in platypus (*Ornithorhynchus anatinus*): Upsilon (υ)-crystallin. *Molecular Biology and Evolution* 20:994–998.

Vitale, F. 2018. *Biodeconstruction: Jacques Derrida and the Life Sciences*. New York: SUNY Press, 2018.

von zur Gathen, J. 2006. Zimmermann telegram: The original draft. *Cryptologia* 31:2–37.

Wachter, A. 2010. Riboswitch-mediated control of gene expression in eukaryotes. *RNA Biology* 7:67–76.

Wachter, A., M. Tunc-Ozdemir, B. C. Grove, P. J. Green, D. K. Shintani, and R. R. Breaker. 2007. Riboswitch control of gene expression in plants by splicing and alternative 3′ end processing of mRNAs. *Plant Cell* 19:3437–3450.

Waddell, D. R. 1982. A predatory slime mould. *Nature* 298:464–466.

Waddington, C. H. 1957. *The Strategy of the Genes*. London: George Allen & Unwin.

Wagner, G. P. 1989. The biological homology concept. *Annual Review of Ecology and Systematics* 20:51–69.

Wagner, G. P. 2014. *Homology, Genes, and Evolutionary Innovation*. Princeton: Princeton University Press.

Wardlaw, C. W. 1952. *Phylogeny and Morphogenesis*. London: Macmillan.

Waters, C. K. 2005. Why genic and multi-level selection theories are here to stay. *Philosophy of Science* 72:311–333.

Waters, C. K. 2007. Causes that make a difference. *Journal of Philosophy* 104:551–579.

Watson, P. Y., and M. J. Fedor. 2011. The *glmS* riboswitch integrates signals from activating and inhibitory metabolites in vivo. *Nature Structural and Molecular Biology* 18:359–363.

Watt, W. B. 2013. Causal mechanisms of evolution and the capacity for niche construction. *Biology & Philosophy* 28:757–766.

Weaver, W. 1949. The mathematics of communication. *Scientific American* 181(1):11–15.

Weismann, A. 1890. Prof. Weismann's theory of heredity. *Nature* 41:317–323.

Weismann, A. 1896. Germinal selection. *Monist* 6:250–293.

Whewell, W. 1833. *Astronomy and General Physics Considered with Reference to Natural Theology*. Third Bridgewater Treatise. Philadelphia: Carey, Lee & Blanchard.

Whewell, W. 1845. *Indications of the Creator*. Philadelphia: Carey & Hart.

White, H. B. 1976. Coenzymes as fossils of an earlier metabolic state. *Journal of Molecular Evolution* 7:101–104.

Wilkins, J. F., and D. Haig. 2001. Genomic imprinting of two antagonistic loci. *Proceedings of the Royal Society B* 268:1861–1867.

Wilkins, J. F., and D. Haig. 2003. What good is genomic imprinting: The function of parent-specific gene expression. *Nature Reviews Genetics* 4:359–368.

Williams, A. F., and A. N. Barclay. 1988. The immunoglobulin superfamily—domains for cell surface recognition. *Annual Review of Immunology* 6:381–405.

Williams, C. A., H. Angelman, J. Clayton-Smith, D. J. Driscoll, J. E. Hendrickson, J. H. M. Knoll, et al. 1995. Angelman syndrome: Consensus for diagnostic criteria. *American Journal of Medical Genetics* 56:237–238.

Williams, C. A., A. L. Beaudet, J. Clayton-Smith, J. H. Knoll, M. Kyllerman, L. A. Laan, et al. 2005. Angelman syndrome 2005: Updated consensus for diagnostic criteria. *American Journal of Medical Genetics* 140A:413–418.

Williams, G. C. 1966. *Adaptation and Natural Selection*. Princeton: Princeton University Press.

Williams, G. C. 1986. Comments on Sober's *The Nature of Selection*. *Biology & Philosophy* 1:114–122.

Williams, G. C. 1992. *Natural Selection: Domains, Levels, and Challenges*. Oxford: Oxford University Press.

Wilson, D. S. 1980). *The Natural Selection of Populations and Communities*. Menlo Park, CA: Benjamin/Cummings.

Wilson, D. S., and E. Sober. 1994. Reintroducing group selection to the human behavioral sciences. *Behavioral and Brain Sciences* 17:585–654.

Winkler, W., A. Nahvi, and R. R. Breaker. 2002. Thiamine derivatives bind messenger RNAs directly to regulate bacterial gene expression. *Nature* 419:952–956.

Winnie, J. A. 2000. Information and structure in molecular biology: Comments on Maynard Smith. *Philosophy of Science* 67:517–526.

Wistow, G. 1993. Lens crystallins: Gene recruitment and evolutionary dynamism. *Trends in Biochemical Sciences* 18:301–306.

Wistow, G. J., J. W. M. Mulders, and W. W. de Jong. 1987. The enzyme lactate dehydrogenase as a structural protein in avian and crocodilian lenses. *Nature* 326:622–624.

Wittgenstein, L. 1958. *Preliminary Studies for the "Physiological Investigations": Generally Known as the Blue and Brown Books*. Oxford: Blackwell.

Wrangham, R. W. 1999. Evolution of coalitionary killing. *Yearbook of Physical Anthropology* 42:1–30.

Wunner, W. H. 2007. Rabies virus. In *Rabies*, 2nd ed., ed. A. C. Jackson and W. H. Wunner, 23–68. Amsterdam: W. H. Elsevier.

Yamada, K. A., and J. J. Volpe. 1990. Angelman's syndrome in infancy. *Developmental Medicine and Child Neurology* 32:1005–1011.

Yu, S., D. Yu, E. Lee, M. Eckhaus, R. Lee, Z. Corria, et al. 1998. Variable and tissue-specific hormone resistance in heterotrimeric G_s protein α-subunit ($G_s\alpha$) knockout mice is due to tissue-specific imprinting of the $G_s\alpha$ gene. *Proceedings of the National Academy of Sciences USA* 95:8715–8720.

Zhivotovsky, L. A., M. W. Feldman, and F. B. Christiansen. 1994. Evolution of recombination among multiple selected loci: A generalized reduction principle. *Proceedings of the National Academy of Sciences USA* 91:1079–1083.

Zwanzig, R., A. Szabo, and B. Bagchi. 1992. Levinthal's paradox. *Proceedings of the National Academy of Sciences USA* 89:20–22.

Zyskind, J. W., and D. W. Smith. 1992. DNA replication, the bacterial cell cycle, and cell growth. *Cell* 69:5–8.

Sources

Chapters 2 through 11 are modified versions of previously published essays.

Chapter 2: Haig, D. 1997. "The Social Gene." In *Behavioural Ecology*, 4th ed., ed. J. R. Krebs and N. B. Davies, 284–304. Oxford: Blackwell Scientific.

Chapter 3: Haig, D. 2006. "The Gene Meme." In *Richard Dawkins: How a Scientist Changed the Way We Think*, ed. A. Grafen and M. Ridley, 50–65. Oxford: Oxford University Press.

Chapter 4: Haig, D. 2012. The strategic gene. *Biology and Philosophy* 27:461–479.

Chapter 5: Haig, D. 2008. "Conflicting Messages: Genomic Imprinting and Internal Communication." In *Sociobiology of Communication*, ed. P. D'Ettorre and D. P. Hughes, 209–223. Oxford: Oxford University Press.

Chapter 6: Haig, D. 2006. "Intrapersonal Conflict." In *Conflict*, ed. M. K. Jones and A. C. Fabian, 8–22. Cambridge: Cambridge University Press, Cambridge.

Chapter 7: Haig, D. 2003. "On Intrapersonal Reciprocity." *Evolution and Human Behavior* 24:418–425.

Chapter 8: Haig, D. 2011. "Sympathy with Adam Smith and Reflexions on Self." *Journal of Economic Behavior and Organization* 77:4–13.

Chapter 9: Haig, D. 2013. "Proximate and Ultimate Causes: How Come? And What For?" *Biology and Philosophy* 28:781–786.

Chapter 10: Haig, D. 2015. "Sameness, Novelty, and Nominal Kinds." *Biology and Philosophy* 30:857–872.

Chapter 11: Haig, D. 2014. "Fighting the Good Cause: Meaning, Purpose, Difference, and Choice." *Biology and Philosophy* 29:675–697.

Index

Page numbers followed by an "f" indicate figures.

Eukaryotes, 51, 52, 84, 216–217.
  See also Slime mold
  clade of, 216
  replication, 37–38
  RNA and, 238, 324, 337
Eukaryotic alliance, 43–45
Eukaryotic cell cycle, 38, 52
Eukaryotic cells, 39, 43, 45
  features of, 38
Eukaryotic genomes, 238
Eukaryotic nucleus, 39, 43
Eukaryotic recombination, 51, 52
Euphemisms, 389
Evil, 17, 125, 356
Evolution
  goal/purpose, 194
  meanings and uses of the
    term, 13, 208
Evolutionary biology, xvi, 126. See
  also specific topics
  adaptationist explanations
    within, 374–375
  criticisms of, 192
  functional biology, functional
    biologists, and, 192–194
  Geisteswissenschaften and,
    371
  Ernst Mayr and, 158, 183,
    192–197
  Adam Smith and, 183
Evolutionary bricolage, 329,
  348
Evolutionary causes. See also
  Ultimate causes
  defined, 201
  Ernst Mayr on, 193–194, 201

  vs. proximate causes, 193–194,
    198, 201
Evolutionary genes, 82–86, 254,
  255
  asexual genotypes as, 258
  chromosomes and, 82–86
  concept of, 82, 85, 86
  defined, 254
  DNA and, 82–83, 85, 254
  gene pools and, 86
  sexual eukaryotes and, 84
Evolutionary narratives, 371. See
  also Narratives
Evolutionary questions
  vs. developmental questions,
    263
  vs. ontogenetic questions, 93
Evolutionary stable strategy (ESS),
  149, 150
Evolutionary synthesis, 13
Evolutionary theory, xix. See also
  specific topics
  causes of disagreement within,
    272, 343
  developmental biology
    and, 14
  mathematics and, xix
Exaptation, 273, 274
  vs. adaptation, 273, 319
  defined, 223
Explanation
  four kinds of, 158
  two domains of, 262
  understanding and, 361–362,
    384–385
Expression, axis of, 262